AF685200

Calcium, Vitamin D and Aging in Humans

Calcium, Vitamin D and Aging in Humans

Guest Editors

Stefano Gonnelli
Ligia J. Dominguez

MDPI

Basel • Beijing • Wuhan • Barcelona • Belgrade • Novi Sad • Cluj • Manchester

Guest Editors
Stefano Gonnelli
Department of Medicine
Surgery and Neuroscience
University of Siena
Siena
Italy

Ligia J. Dominguez
Department of Medicine
University of Palermo
Palermo
Italy

Editorial Office
MDPI AG
Grosspeteranlage 5
4052 Basel, Switzerland

This is a reprint of the Special Issue, published open access by the journal *Nutrients* (ISSN 2072-6643), freely accessible at: https://www.mdpi.com/journal/nutrients/special_issues/2638CC95TL.

For citation purposes, cite each article independently as indicated on the article page online and as indicated below:

Lastname, A.A.; Lastname, B.B. Article Title. *Journal Name* **Year**, *Volume Number*, Page Range.

ISBN 978-3-7258-6748-6 (Hbk)
ISBN 978-3-7258-6749-3 (PDF)
https://doi.org/10.3390/books978-3-7258-6749-3

© 2026 by the authors. Articles in this reprint are Open Access and distributed under the Creative Commons Attribution (CC BY) license. The reprint as a whole is distributed by MDPI under the terms and conditions of the Creative Commons Attribution-NonCommercial-NoDerivs (CC BY-NC-ND) license (https://creativecommons.org/licenses/by-nc-nd/4.0/).

Contents

About the Editors

Stefano Gonnelli

Stefano Gonnelli was born in Siena (Italy) and graduated in Medicine and Surgery from the University of Siena in 1978. He obtained specializations in Cardiology, Endocrinology, and Preventive Medicine, and in 1979–1980, he completed his military service as a Medical Officer. In 1981, he joined the University of Siena, where he spent his entire academic career, first as a Researcher, then as an Associate Professor, and finally as a Full Professor in Internal Medicine. Director of the Postgraduate School in Emergency and Urgent Medicine at the University of Siena from 2012 to 2019 and of the Postgraduate School of in Internal Medicine from 2019 to 2023; he is currently a faculty member in the Postgraduate Schools in Internal Medicine and Geriatrics. He was President of the Italian Society of Osteoporosis, Mineral Metabolism, and Skeletal Diseases from 2017 to 2019. His scientific activity has mainly focused on vitamin D and metabolic bone diseases, with particular emphasis on instrumental and laboratory diagnostic methods. He has authored over 250 publications in national and international peer-reviewed scientific journals and has been invited as a speaker at numerous national and international meetings.

Ligia J. Dominguez

Ligia J. Dominguez was born in Colombia where she graduated in medicine and completed her residency in Internal Medicine at the "Universidad del Rosario" in Bogota. She then joined the Division of Endocrinology, Metabolism and Hypertension at Wayne State University (Detroit, MI) carrying out basic and clinical research, and completing the specialization in Endocrinology. She moved to Italy in 1996, continuing her clinical and research interests in the fields of diabetes and its complications, nutrition, and bone and mineral metabolism, in relation to aging. Prof. Dominguez is currently a full professor at the University of Palermo (Italy), where she teaches in the Faculty of Medicine, Nursing, Sports Sciences, and in the Postgraduate School of Geriatrics and Internal Medicine. She has authored over 300 publications in peer-reviewed national and international scientific journals and has been invited as speaker at national and international meetings.

Preface

There is growing recognition that nutrition plays a key role in age-related chronic diseases and mortality. Calcium and vitamin D are particularly important for musculoskeletal health, and deficiencies increase the risk of osteoporosis and fragility fractures in older adults. Supplementation, especially in frail and institutionalized individuals, reduces hip and fracture risk and may improve muscle function and lower fall risk. Beyond skeletal effects, low vitamin D levels have been associated with higher risks of dementia, diabetes, cancer, and infections, although further well-designed studies are needed. This Special Issue highlights recent evidence on the role of calcium and vitamin D in health, particularly during aging.

Stefano Gonnelli and Ligia J. Dominguez
Guest Editors

nutrients MDPI

Editorial

Calcium, Vitamin D, and Aging in Humans

Ligia J. Dominguez [1,*] and Stefano Gonnelli [2]

1 Department of Medicine and Surgery, "Kore" University of Enna, 94100 Enna, Italy
2 Section of Internal Medicine, Department of Medicine, Surgery and Neuroscience, University of Siena, 53100 Siena, Italy; stefano.gonnelli@unisi.it
* Correspondence: ligia.dominguez@unikore.it

There is currently a growing awareness that nutritional factors have major impacts on the risk of age-associated chronic non-communicable diseases and mortality [1]. Two of these nutrients, calcium and vitamin D, have been particularly implicated in maintaining musculoskeletal health, which are increasingly common conditions at different stages of life but particularly in old age. In fact, adults and older persons with inadequate dietary calcium intake or insufficient vitamin D levels have an increased risk of osteoporosis and fragility fractures, which are continuously increasing worldwide [2]. There is evidence that the supplementation of these nutrients in frail older adults living in long-term care facilities reduces the risk of hip and other fragility fractures [3–5]. In addition, vitamin D supplementations may improve muscle function and contribute to decreasing the risks of falling in older adults with deficiency of this nutrient [6–8]. Moreover, vitamin D may have biological effects well beyond the skeleton, and several studies support an association between low vitamin D concentrations and an increased risk of dementia, diabetes, cancer, and infectious diseases [9–11]. However, further well-designed studies are needed to better define the role of vitamin D in the prevention and treatment of these diseases in older adults. This Special Issue brings together some articles with recent and relevant data on these two nutrients that are essential for human health at all ages, and, in particular during aging.

One of the articles explored the efficacy of the recommendations promoted for a decade in Poland aiming to prevent vitamin D deficiency in older adults [12]. Lech et al. analyzed data from community-dwelling participants of the PolSenior 2 study who were admitted to a geriatrics ward to evaluate the prevalence of vitamin D deficiency and the adherence to recommended health-promoting behaviors (i.e., taking vitamin D supplements). The authors found that among 240 participants, over two-thirds (67.5%) had inadequacy of 25OHD levels (<30 ng/mL), and 15% had profound deficiency (<10 ng/mL) of this fundamental nutrient. Only 18.3% of participants were on vitamin D supplements before admission, and, as expected, insufficiency was significantly more prevalent in those without using supplements (70.9% vs. 52.3%) and in obese participants (body mass index [BMI] ≥ 30 kg/m^2). Furthermore, obese participants had over twice the risk of vitamin D deficiency compared to those not receiving supplements. Thus, despite recommendations, the prevalence of vitamin D deficiency is remarkably high in community-dwelling Polish older adults, and targeted interventions and education for older adults, caregivers, and health providers are still needed to fill this gap. Currently, it could be hypothesized that vitamin D deficiency in children no longer exists in Western countries. As a comparison with what occurs in adulthood and in old age, the study by Van de Walle and coworkers evaluated the vitamin D status among a large sample (n = 14,887) of Belgian children aged 0–18 years between 2014 and 2021 [13]. Even in this young group, 17.7% were vitamin D severely deficient (<12 ng/mL), 25.2% were insufficient (12–20 ng/mL), and another large proportion (28.3%) was borderline (20–30 ng/mL). Sufficient levels (>30 ng/mL) were met in only 28.8% of the sample. Adolescents aged 13–18 years had the highest prevalence of severe vitamin D deficiency (24.9%). Thus, vitamin D deficiency and insufficiency are also remarkably prevalent in this young age group. The authors warn that Belgian health authorities should be aware of these results, as the current Belgian recommendation suggests ceasing vitamin D supplementation at the age of six. These children, who will become adults and old people in the future, could be exposed to a lifelong vitamin D deficiency with worrying consequences for their health in later life.

Cianferotti and coworkers reviewed available randomized controlled trials, prospective and intervention studies, and systematic reviews on the screening, assessment, and management of nutritional and vitamin D status and calcium intake in patients before and after having a hip fracture [14]. The in-depth review of the studies available in recent years, with all their contradictions, collectively suggests that targeted nutritional interventions, including calcium and protein intake through diet or supplements, together with rehabilitation programs, can significantly contribute to improved recovery and mobility after hip fractures. Such strategies can also contribute to the prevention of future new fractures by improving overall bone and skeletal muscle health. The review also emphasizes the role of fracture liaison services, which represent the most reliable model of management for hip fracture patients.

The most widely used supplements of vitamin D are ergocalciferol, cholecalciferol, and calcifediol, with differences in their utilization in European and American countries in part depending on the availability of these compounds, i.e., calcifediol is not available in many countries. Calcifediol is formed by the addition of a 25-hydroxyl group to its precursor, mainly by the enzymes microsomal CYP2R1 and mitochondrial CYP27A1 in the liver [9,15]. Previous studies have shown that calcifediol administration determines a faster increase in circulating 25OHD when compared to cholecalciferol [16,17]. In fact, calcifediol is rapidly absorbed by intestinal cells and transported through the portal vein, hence becoming immediately available in the circulation system. In order to test the efficacy and safety of monthly 0.266 mg calcifediol, Occhiuto et al. performed an open-label investigation with calcifediol administration for two years in post-menopausal women with hypovitaminosis D [18]. The results of this study showed a significant continuous increase in mean 25OHD values even if the mean values at month 24 were not significantly different vs. those at 12 months, but in any case, stable. Similar results were observed for the reduction in mean bone alkaline phosphatase, while no safety concerns were reported. It is worth mentioning that this is the first long-term study with calcifediol administration showing stable and sustained 25OHD concentrations compared to previous investigations that had a much shorter duration, i.e., a few months [19,20].

In addition to the recognized functions in the regulation of calcium absorption, bone remodeling, and bone growth, vitamin D plays a fundamental role as a hormone due to its involvement in numerous enzymatic, physiological, metabolic, and pathophysiological processes in various human organs and systems. There is increasing evidence supporting the role of vitamin D in pancreatic islet dysfunction and insulin resistance in type 2 diabetes, with recent convincing data of the true role of vitamin D supplementation in the prevention of incident type 2 diabetes. However, most meta-analyses evaluating this role have been conducted in people aged 50–60 years [21–24], with only one focusing on older adults, even if this is the population at highest risk of hypovitaminosis D and type 2 diabetes. Therefore, we conducted an update of the previous systematic review and meta-analysis examining whether hypovitaminosis D could predict incident diabetes in prospective longitudinal studies including older adults [25]. We found that low 25OHD levels were associated with higher incident type 2 diabetes in older adults even after adjusting for relevant potential confounders, confirming and updating the previous meta-analysis conducted in 2017 [26].

Various studies have investigated the association of dietary components and patterns with osteoporosis and bone mineral density in different ethnic groups. However, there are limited studies conducted in Arabian populations. In order to fill this gap, the study by Al-Daghri and collaborators examined whether dietary patterns, and calcium intake in particular, were associated with osteoporosis risk in Saudi Arabian adults [27]. The authors examined cross-sectional data from almost two thousand patients (416 men and

1534 women) with known risk factors for osteoporosis from those attending tertiary hospitals in Riyadh City. The prevalence of osteoporosis was 21.3% and, as expected, was more frequent in women (93.5%) and older participants with lower BMI. The mean calcium intake was only 445.1 mg/day (vs. recommended dietary intake of 1300 mg/day). The consumption of tea was associated with a 20% reduced presence of osteoporosis, while the consumption of fish and eggs was associated with a 10% lower presence of this condition. Conversely, the consumption of biscuits, cakes, and bread slices was significantly associated with a 30% higher presence of osteoporosis. The authors conclude that the finding of extremely low dietary calcium intake among Saudi adults already at risk of osteoporosis suggests an urgent need to promote correction of this alarming dietary deficit in the Saudi adult population.

Aging is a complex and multifactorial process resulting from a number of intrinsic factors and their interactions with the external environment (e.g., air pollutants and radiation, ultraviolet sunlight, physical exercise, diet quality, chemicals in the water), which have become an extensive field of study. In 2013, nine hallmarks of aging were proposed by Lopez-Otin et al., comprising genomic instability, telomere attrition, epigenetic alterations, mitochondrial dysfunction, loss of proteostasis, deregulated nutrient sensing, cellular senescence, stem cell exhaustion, and altered intercellular communication [28], which were expanded in 2023, adding autophagy, microbiome disturbance, and inflammation, among other emerging ones [29,30]. This complex network of cellular and intercellular connections might represent possible targets for the research of agents with pleiotropic effects. Vitamin D has a positive impact not only on skeletal muscle and bone health but also on multiple extra-skeletal organs that express the vitamin D receptor. Thus, vitamin D and its receptor could be molecules potentially targeting the hallmarks of the aging network. In a comprehensive review article, Ruggiero et al. [31] report evidence on the potential effects of vitamin D on the hallmarks of aging, which are still underrepresented in humans and mainly based on pre-clinical models. However, studies in humans seem to confirm the modulatory effect of vitamin D on some of the hallmarks of aging and diseases, opening up new possibilities for future research in this fascinating and useful field.

It is well known that the synthesis of vitamin D_3 in the human skin is initiated by solar ultraviolet radiation (UVR) exposure of the precursor 7-dehydrocholesterol (7DHC), but the influence of age on this stage of vitamin D metabolism is not completely clear. Borecka et al. performed a prospective standardized study in a group of healthy adults aged $\geq$ 65 and $\leq$40 years examining baseline skin 7DHC concentrations and the impact of older age on serum vitamin D_3 response to solar simulated low-dose UVR to ~35% of the body surface area [32]. The concentration of 7DHC measured in skin biopsies was from unexposed skin (baseline), as well as immediately after UVR exposure and 24 h post-UVR exposure. Blood 25OHD assays were performed at baseline, 24 h, and 7 d post-UVR. Baseline skin 7DHC were similar and not significantly different in younger vs. older adults. Baseline serum 25OHD was also similar in younger vs. older adults and showed significant increases in both groups post-UVR without significant differences between the age groups. Thus, skin 7DHC concentration was not a limiting factor for vitamin D_3 production in older adults relative to younger adults. These results may help public health guidance on sun exposure and vitamin D nutrition, with particular relevance to the growing populations of healthy older adults. This is in contrast to a previous study of about 40 years ago suggesting that low vitamin D status in older adults was at least partially attributable to lower levels of the skin precursor 7DHC [33]. The authors admit that the main limitation of this study is the relatively low number of volunteers, as intensive in vivo multiple biopsy studies are challenging to recruit a higher number of participants, and this potentially resulted in missed detection of significant differences. Furthermore, the participants were healthy, which prevents the results from being applied to frail older participants with multimorbidity and polypharmacy. Nevertheless, the results are novel and encourage future larger studies to confirm these intriguing results.

Thus, the articles of this Special Issue add important information on the role of two key nutrients, calcium and vitamin D, in human aging and the importance of identifying their deficiency/insufficiency with the aim of implementing an appropriate correction of these deficits.

Author Contributions: Writing—original draft preparation, L.J.D.; writing—review and editing, S.G. All authors have read and agreed to the published version of the manuscript.

Funding: This article received no external funding.

Conflicts of Interest: The authors declare no conflicts of interest.

References

1. Ruthsatz, M.; Candeias, V. Non-communicable disease prevention, nutrition and aging. *Acta Biomed.* **2020**, *91*, 379–388. [PubMed]
2. Collaborators, G.B.D.F. Global, regional, and national burden of bone fractures in 204 countries and territories, 1990-2019: A systematic analysis from the Global Burden of Disease Study 2019. *Lancet Healthy Longev.* **2021**, *2*, e580–e592.
3. Chapuy, M.C.; Arlot, M.E.; Duboeuf, F.; Brun, J.; Crouzet, B.; Arnaud, S.; Delmas, P.D.; Meunier, P.J. Vitamin D3 and calcium to prevent hip fractures in elderly women. *N. Engl. J. Med.* **1992**, *327*, 1637–1642. [CrossRef] [PubMed]
4. Chapuy, M.C.; Pamphile, R.; Paris, E.; Kempf, C.; Schlichting, M.; Arnaud, S.; Garnero, P.; Meunier, P.J. Combined calcium and vitamin D3 supplementation in elderly women: Confirmation of reversal of secondary hyperparathyroidism and hip fracture risk: The Decalyos II study. *Osteoporos. Int.* **2002**, *13*, 257–264. [CrossRef] [PubMed]
5. Yao, P.; Bennett, D.; Mafham, M.; Lin, X.; Chen, Z.; Armitage, J.; Clarke, R. Vitamin D and Calcium for the Prevention of Fracture: A Systematic Review and Meta-analysis. *JAMA Netw. Open* **2019**, *2*, e1917789. [CrossRef]
6. Barbagallo, M.; Veronese, N.; Di Prazza, A.; Pollicino, F.; Carruba, L.; La Carrubba, A.; Dominguez, L.J. Effect of Calcifediol on Physical Performance and Muscle Strength Parameters: A Systematic Review and Meta-Analysis. *Nutrients* **2022**, *14*, 1860. [CrossRef]
7. Tan, L.; He, R.; Zheng, X. Effect of vitamin D, calcium, or combined supplementation on fall prevention: A systematic review and updated network meta-analysis. *BMC Geriatr.* **2024**, *24*, 390. [CrossRef]
8. Thanapluetiwong, S.; Chewcharat, A.; Takkavatakarn, K.; Praditpornsilpa, K.; Eiam-Ong, S.; Susantitaphong, P. Vitamin D supplement on prevention of fall and fracture: A Meta-analysis of Randomized Controlled Trials. *Medicine* **2020**, *99*, e21506. [CrossRef]
9. Dominguez, L.J.; Farruggia, M.; Veronese, N.; Barbagallo, M. Vitamin D Sources, Metabolism, and Deficiency: Available Compounds and Guidelines for Its Treatment. *Metabolites* **2021**, *11*, 255. [CrossRef]
10. Passeri, G.; Giannini, S. Benefits of Vitamin D in Health and Diseases. *Nutrients* **2023**, *15*, 2419. [CrossRef]
11. Charoenngam, N.; Holick, M.F. Immunologic Effects of Vitamin D on Human Health and Disease. *Nutrients* **2020**, *12*, 2097. [CrossRef] [PubMed]
12. Lech, M.A.; Warpechowski, M.; Wojszel, A.; Rentflejsz, J.; Swietek, M.; Wojszel, Z.B. Vitamin D Status among Patients Admitted to a Geriatric Ward-Are Recommendations for Preventing Its Deficiency Effective Enough? *Nutrients* **2024**, *16*, 193. [CrossRef] [PubMed]
13. Van de Walle, L.; Vandenplas, Y.; Toelen, J.; Raaijmakers, A. Vitamin D Status in Belgian Children: A Regional Study. *Nutrients* **2024**, *16*, 657. [CrossRef] [PubMed]
14. Cianferotti, L.; Bifolco, G.; Caffarelli, C.; Mazziotti, G.; Migliaccio, S.; Napoli, N.; Ruggiero, C.; Cipriani, C. Nutrition, Vitamin D, and Calcium in Elderly Patients before and after a Hip Fracture and Their Impact on the Musculoskeletal System: A Narrative Review. *Nutrients* **2024**, *16*, 1773. [CrossRef] [PubMed]
15. Minisola, S.; De Martino, V. Disentanglement among vitamins D. *Gynecol. Endocrinol.* **2022**, *38*, 97–98. [CrossRef]
16. Bouillon, R.; Quesada Gomez, J.M. Comparison of calcifediol with vitamin D for prevention or cure of vitamin D deficiency. *J. Steroid Biochem. Mol. Biol.* **2023**, *228*, 106248. [CrossRef]
17. Charoenngam, N.; Kalajian, T.A.; Shirvani, A.; Yoon, G.H.; Desai, S.; McCarthy, A.; Apovian, C.M.; Holick, M.F. A pilot-randomized, double-blind crossover trial to evaluate the pharmacokinetics of orally administered 25-hydroxyvitamin D3 and vitamin D3 in healthy adults with differing BMI and in adults with intestinal malabsorption. *Am. J. Clin. Nutr.* **2021**, *114*, 1189–1199. [CrossRef]
18. Occhiuto, M.; Pepe, J.; Colangelo, L.; Lucarelli, M.; Angeloni, A.; Nieddu, L.; De Martino, V.; Minisola, S.; Cipriani, C. Effect of 2 Years of Monthly Calcifediol Administration in Postmenopausal Women with Vitamin D Insufficiency. *Nutrients* **2024**, *16*, 1754. [CrossRef]
19. Bischoff-Ferrari, H.A.; Dawson-Hughes, B.; Stocklin, E.; Sidelnikov, E.; Willett, W.C.; Edel, J.O.; Stahelin, H.B.; Wolfram, S.; Jetter, A.; Schwager, J.; et al. Oral supplementation with 25(OH)D3 versus vitamin D3: Effects on 25(OH)D levels, lower extremity function, blood pressure, and markers of innate immunity. *J. Bone Miner. Res.* **2012**, *27*, 160–169. [CrossRef]
20. Russo, S.; Carlucci, L.; Cipriani, C.; Ragno, A.; Piemonte, S.; Fiacco, R.D.; Pepe, J.; Fassino, V.; Arima, S.; Romagnoli, E.; et al. Metabolic changes following 500 mug monthly administration of calcidiol: A study in normal females. *Calcif. Tissue Int.* **2011**, *89*, 252–257. [CrossRef]

21. Song, Y.; Wang, L.; Pittas, A.G.; Del Gobbo, L.C.; Zhang, C.; Manson, J.E.; Hu, F.B. Blood 25-hydroxy vitamin D levels and incident type 2 diabetes: A meta-analysis of prospective studies. *Diabetes Care* **2013**, *36*, 1422–1428. [CrossRef] [PubMed]
22. Barbarawi, M.; Zayed, Y.; Barbarawi, O.; Bala, A.; Alabdouh, A.; Gakhal, I.; Rizk, F.; Alkasasbeh, M.; Bachuwa, G.; Manson, J.E. Effect of Vitamin D Supplementation on the Incidence of Diabetes Mellitus. *J. Clin. Endocrinol. Metab.* **2020**, *105*, 2857–2868. [CrossRef] [PubMed]
23. Zhang, Y.; Tan, H.; Tang, J.; Li, J.; Chong, W.; Hai, Y.; Feng, Y.; Lunsford, L.D.; Xu, P.; Jia, D.; et al. Effects of Vitamin D Supplementation on Prevention of Type 2 Diabetes in Patients With Prediabetes: A Systematic Review and Meta-analysis. *Diabetes Care* **2020**, *43*, 1650–1658. [CrossRef] [PubMed]
24. Pittas, A.G.; Kawahara, T.; Jorde, R.; Dawson-Hughes, B.; Balk, E.M. Vitamin D and Risk for Type 2 Diabetes in People With Prediabetes. *Ann. Intern. Med.* **2023**, *176*, eL230202. [CrossRef]
25. Dominguez, L.J.; Veronese, N.; Marrone, E.; Di Palermo, C.; Iommi, C.; Ruggirello, R.; Caffarelli, C.; Gonnelli, S.; Barbagallo, M. Vitamin D and Risk of Incident Type 2 Diabetes in Older Adults: An Updated Systematic Review and Meta-Analysis. *Nutrients* **2024**, *16*, 1561. [CrossRef]
26. Lucato, P.; Solmi, M.; Maggi, S.; Bertocco, A.; Bano, G.; Trevisan, C.; Manzato, E.; Sergi, G.; Schofield, P.; Kouidrat, Y.; et al. Low vitamin D levels increase the risk of type 2 diabetes in older adults: A systematic review and meta-analysis. *Maturitas* **2017**, *100*, 8–15. [CrossRef]
27. Al-Daghri, N.M.; Hussain, S.D.; Alnaami, A.M.; Aljohani, N.; Sabico, S. Dietary Calcium Intake and Osteoporosis Risk in Arab Adults. *Nutrients* **2023**, *15*, 2829. [CrossRef]
28. Lopez-Otin, C.; Blasco, M.A.; Partridge, L.; Serrano, M.; Kroemer, G. The hallmarks of aging. *Cell* **2013**, *153*, 1194–1217. [CrossRef]
29. Schmauck-Medina, T.; Moliere, A.; Lautrup, S.; Zhang, J.; Chlopicki, S.; Madsen, H.B.; Cao, S.; Soendenbroe, C.; Mansell, E.; Vestergaard, M.B.; et al. New hallmarks of ageing: A 2022 Copenhagen ageing meeting summary. *Aging* **2022**, *14*, 6829–6839. [CrossRef]
30. Lopez-Otin, C.; Blasco, M.A.; Partridge, L.; Serrano, M.; Kroemer, G. Hallmarks of aging: An expanding universe. *Cell* **2023**, *186*, 243–278. [CrossRef]
31. Ruggiero, C.; Tafaro, L.; Cianferotti, L.; Tramontana, F.; Macchione, I.G.; Caffarelli, C.; Virdis, A.; Ferracci, M.; Rinonapoli, G.; Mecocci, P.; et al. Targeting the Hallmarks of Aging with Vitamin D: Starting to Decode the Myth. *Nutrients* **2024**, *16*, 906. [CrossRef] [PubMed]
32. Borecka, O.; Dutton, J.J.; Tang, J.C.Y.; Fraser, W.D.; Webb, A.R.; Rhodes, L.E. Comparative Study of Healthy Older and Younger Adults Shows They Have the Same Skin Concentration of Vitamin D(3) Precursor, 7-Dehydrocholesterol, and Similar Response to UVR. *Nutrients* **2024**, *16*, 1147. [CrossRef] [PubMed]
33. MacLaughlin, J.; Holick, M.F. Aging decreases the capacity of human skin to produce vitamin D3. *J. Clin. Investig.* **1985**, *76*, 1536–1538. [CrossRef] [PubMed]

Disclaimer/Publisher's Note: The statements, opinions and data contained in all publications are solely those of the individual author(s) and contributor(s) and not of MDPI and/or the editor(s). MDPI and/or the editor(s) disclaim responsibility for any injury to people or property resulting from any ideas, methods, instructions or products referred to in the content.

nutrients

MDPI

Article

Dietary Calcium Intake and Osteoporosis Risk in Arab Adults

Nasser M. Al-Daghri [1,*], Syed Danish Hussain [1], Abdullah M. Alnaami [1], Naji Aljohani [2,3] and Shaun Sabico [1]

[1] Chair for Biomarkers of Chronic Diseases, Biochemistry Department, College of Science, King Saud University, Riyadh 11451, Saudi Arabia
[2] Obesity, Endocrine and Metabolism Center, King Fahad Medical City, Riyadh 12231, Saudi Arabia
[3] College of Medicine, Alfaisal University, Riyadh 11533, Saudi Arabia
* Correspondence: ndaghri@ksu.edu.sa; Tel.: +966-114-675-939; Fax: +966-114-675-931

Abstract: Osteoporosis is a major public health concern in Saudi Arabia's aging population. There is particularly limited information on how diet affects bone loss in this ethnic group. The purpose of this study was to examine the association between dietary calcium (Ca) intake and osteoporosis risk in Saudi adults. A total of 1950 patients (416 males and 1534 females) with known risk factors for osteoporosis participated in this cross-sectional study. A short questionnaire (CaQ) was used to assess dietary Ca intakes in patients attending tertiary hospitals in Riyadh City. The prevalence of osteoporosis was 21.3% and was more common in females (93.5%). Patients with osteoporosis were older ($p < 0.001$) and had lower BMI ($p < 0.001$). Results showed that the overall mean Ca intake was only 445.1 mg/day (recommended dietary intake of 1300 mg/day). Tea intake (OR = 0.8 95%CI: 0.7–1.0; $p = 0.02$) and consumption of fish and eggs (OR = 0.9 95%CI: 0.8–1.0; $p = 0.01$) were significantly associated with a lower risk of osteoporosis. Furthermore, consumption of biscuits, cake and bread slices were significantly associated with higher incidence of osteoporosis (OR = 1.3 95%CI: 1.0–1.5; $p = 0.02$). In conclusion, extremely low dietary Ca intake was observed among Saudi adults already at risk of osteoporosis. A balanced diet including high amount of Ca, vitamin D and omega-3 fatty acids accompanied by limiting consumption of foods high in saturated fats and glycemic index may be helpful in reducing osteoporosis risk in the Saudi adult population.

Keywords: osteoporosis; dietary intake; bone mineral density

1. Introduction

Osteoporosis is a common disease in Saudi Arabia (SA), resulting from decreased bone mass, strength and microarchitecture, which ultimately elevates the risk of fractures [1]. The World Health Organization (WHO) defined osteoporosis as a decrease in bone mineral density (BMD), measured by dual-energy X-ray absorptiometry (DXA), to at least 2.5 standard deviations below the mean peak BMD of a young adult woman [2]. In SA, the most recent hospital-based study showed that the prevalence of osteoporosis was as high as 64% in men and 58% in post-menopausal women above 50 years old [3]. Factors contributing to the high prevalence of osteoporosis in SA include advancing age, history of previous fracture, family history of low bone mass, menopause and use of corticosteroids [4].

Individuals diagnosed with osteoporosis have a 40% risk of fractures, which can result in mobility impairment and significantly reduce quality of life [1]. Available evidence suggests that bone loss related to age and/or post-menopause is influenced by the level of peak bone mass attained during childhood and adolescence, as well as the rate at which bone is lost [5]. Genetic factors can also explain between 60% and 80% of the variations in peak bone mass, with the remainder modulated by external factors such as nutrition [5–7]. In fact, consuming sufficient dietary calcium (Ca) and protein are necessary to attain ideal peak bone mass and prevent bone loss in adults [8].

Diet plays a significant role in the prevention and management of osteoporosis [8]. Consuming a diet that is rich in Ca, vitamin D, vitamin K and magnesium, and a balanced intake of protein, fruits, milk, yogurt, cheese, dark leafy green vegetables, such as spinach and kale, fish such as sardines, eggs and fortified foods such as breakfast cereals can contribute to good bone health by promoting bone formation and reducing bone loss [9–13]. However, cakes, biscuits and other foods that are high in added sugars and refined grains do not provide the same benefits for bone health and may be associated with an increased risk of inflammation and bone loss over time [7–9].

Several studies have investigated the association of dietary patterns with osteoporosis and BMD in different ethnic groups and have reported mixed results [14–16]. In contrast, there are limited studies conducted in Arabian populations, particularly in SA, as the available literature has focused mainly on osteopenia in post-menopausal women, extra-skeletal effects of vitamin D and more recently, biochemical osteomalacia in Arab adolescents [17–19]. In order to fill this gap, this study examined whether dietary patterns, and Ca intake in particular, are associated with osteoporosis risk in Saudi Arabian adults.

2. Materials and Methods

This multi-center cross-sectional study was taken from the Osteoporosis Registry database of the Chair for Biomarkers of Chronic Diseases (CBCD) in King Saud University (KSU), SA, in collaboration with the Ministry of Health. The project received approval from the Ethics Committee at the College of Medicine, KSU, Riyadh, SA (number H-01-R-012). In brief, 2185 records of all consenting Saudi adults visiting the major tertiary hospitals in Riyadh (King Fahad Medical City (KFMC), King Khalid University Hospital (KKUH) and King Salman Hospital (KSH)) who were advised to have their BMD assessed from 2013–2016 were included [20].

2.1. Study Design and Participants

Data collection for the Osteoporosis Registry started in 2013 and ended in August 2016. Screening of participants for inclusion in the registry were overseen by trained physicians in their respective radiology departments in participating hospitals where a BMD (DXA) scan is available. The inclusion criteria were those preferably, but not limited to, post-menopausal women, those with history of fracture in the last 5 years, diminished height in the last 2 years, men and women with history of malabsorption and nutritional deficiencies (vitamin D and Ca), who had endocrine disorders affecting skeletal mass (e.g., hyperparathyroidism), were at risk for bone density loss (family history of osteoporosis) and had an available BMD scan. All participants were asked for their consent before inclusion in the registry. Participants were excluded if they were <35 years of age or if they have any malignancy, cardiac disorders, lung disease or any disease that required immediate medical attention. For this particular study, participants were excluded if there were no dietary data available. Figure 1 shows the flowchart of participants.

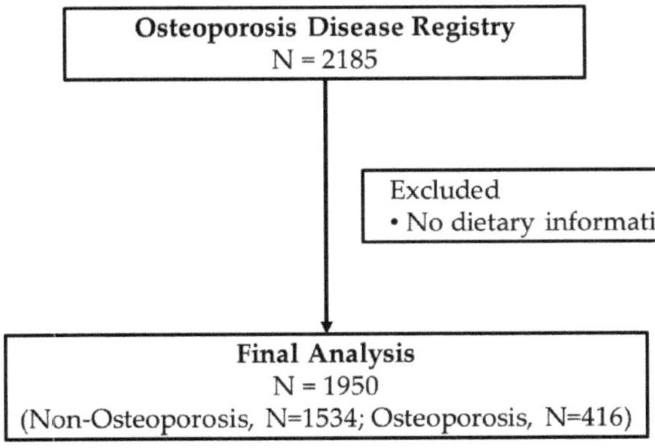

Figure 1. Flowchart of participants.

2.2. Diagnosis of Osteoporosis

Diagnosis of osteoporosis was undertaken by measuring BMD through a DXA scan of the femoral neck or spine and calculating the T-score. A patient with a T-score of ≤ -2.5 (standard deviations, SD, from the mean value of young normal Caucasian women) is classified as having osteoporosis [21,22]. DXA scans were overseen by a qualified technician certified in densitometry, while the machine itself was calibrated using a standard phantom provided by the manufacturer. Upon completion of the test, the BMD findings and corresponding T-scores were forwarded to the Chair for Biomarkers in Chronic Diseases (CBCD) at King Saud University, SA, for classification according to osteoporosis status and data entry.

2.3. Anthropometry, Blood Collection and Sample Analysis

Clinical data were obtained by trained nurses using standard techniques to measure anthropometrics which included height (cm), waist/hip measurements (cm) and weight (kg). Body mass index (BMI) was computed as weight (in kilograms) divided by height (in squared meters). Blood pressure (mmHg) was recorded twice, with a 10-min interval between readings, using a mercurial sphygmomanometer, and the average reading was noted. Fasting blood samples were extracted and subsequently sent to the CBCD laboratory for biochemical analysis using an automated biochemical analyzer (Konelab 20, Thermo-Fischer Scientific, Espoo, Finland). This analysis included serum lipid profiling as well as quantification of fasting glucose, Ca and albumin levels.

2.4. Dietary Intake

Participants were invited for the administration of a short questionnaire (CaQ) adopted from Macdonald et al. [23]. The CaQ is a validated instrument to assess daily dietary Ca intakes [23]. The information collected was on weekly consumption of food and beverages that are common sources of Ca in SA which included milk, tea with milk, coffee with milk, bread, biscuits, cakes, eggs, fish, cheese, yogurt, green vegetables, porridge, pancakes and lasagna. The participants were asked to provide weekly standard servings of the mentioned foods and beverages. Then, weekly Ca intake was calculated by multiplying the amount of Ca present in the standard serving of the food/beverage.

2.5. Data Analysis

Data were analyzed using SPSS, version 21.0. The continuous variables were expressed as the mean with the standard deviation or as the median with the quartile range. The categorical variables were presented as frequencies and percentages. To determine significance, the independent sample *t*-test and Mann–Whitney *U* test were used for normal and non-normal variables, respectively. Age and BMI were adjusted for using univariate analysis of variance. Prior to conducting parametric testing, non-normal variables were log-transformed. Multivariate logistic regression analysis was used to obtain both unadjusted and adjusted odds ratios (OR). A correlation matrix was performed among different sources of dietary Ca prior to univariate modelling to confirm absence of multicollinearity. Significance was set at $p < 0.05$.

3. Results

A total of 1950 patients (N = 1534 (389 males and 1145 females) without osteoporosis and N = 416 (27 males and 389 females with osteoporosis)) were included in this observational study. The prevalence of osteoporosis was 21.3%, of which 93.5% were females. Patients with osteoporosis were significantly older ($p < 0.001$) and had lower BMI ($p < 0.001$). Table 1 shows the anthropometrics, clinical characteristics and biochemical parameters according to osteoporosis status.

Table 1. Anthropometrics and biochemical parameters according to osteoporosis status.

Parameters	Non-Osteoporosis	Osteoporosis	*p*-Value	*p*-Value *
N	1534	416		
Age (years)	53.3 ± 11.5	59.3 ± 8.7	<0.001	--
BMI (kg/m^2)	32.3 ± 6.2	30.3 ± 6.1	<0.001	--
M/F	389/1145	27/389	<0.001	--
T2DM	725 (47.3)	193 (46.4)	0.30	0.06
Hypertension	547 (35.7)	128 (30.8)	0.06	0.01
WHR	0.9 ± 0.4	0.9 ± 0.2	0.40	0.26
Systolic BP (mmHg)	125.3 ± 16.4	124.4 ± 17.8	0.36	0.05
Diastolic BP (mmHg)	77.5 ± 10.0	76.2 ± 10.3	0.02	0.005
Total Cholesterol (mmol/L)	5.1 ± 1.2	5.1 ± 1.1	0.64	0.57
Glucose (mmol/L)	6.4 (5.3–9.5)	6.2 (5.3–8.8)	0.76	0.01
HDL-C (mmol/L)	1.1 ± 0.4	1.2 ± 0.4	0.18	0.12
Triglycerides (mmol/L)	1.6 (1.2–2.2)	1.5 (1.1–2.0)	0.006	0.001
25(OH) D (nmol/L)	59.4 (36.9–89.1)	72.2 (44.8–100.1)	<0.001	0.09
Ca (mmol/L)	2.3 ± 0.3	2.3 ± 0.3	0.16	0.49
Phosphorous (mmol/L)	1.3 (1.0–1.7)	1.2 (1.0–1.7)	0.61	0.50
BMD Spine	1.0 ± 0.3	0.8 ± 0.2	<0.001	<0.001
BMD Femur	1.0 ± 0.1	0.9 ± 0.2	<0.001	<0.001
T-score Spine	−0.8 ± 1.2	−3.1 ± 0.6	<0.001	<0.001
T-score Femur	−0.1 ± 2.9	−1.3 ± 0.9	<0.001	<0.001
Tea with Milk (cups/wk)	0.8 ± 2.9	0.7 ± 3.3	<0.001	0.53
Coffee with milk (cups/wk)	0.5 ± 2.9	0.2 ± 1.0	<0.001	0.12
Slices (Count/wk)	5.6 ± 7.5	6.5 ± 6.9	<0.001	0.42
Biscuits (Count/wk)	0.9 ± 2.9	0.7 ± 2.2	0.06	0.92
Cake (Portion/wk)	0.8 ± 4.7	0.6 ± 3.0	0.001	0.33
Pancakes (Count/wk)	0.3 ± 3.3	0.2 ± 0.7	0.40	0.99
Lasagna (Portion/wk)	1.0 ± 11.2	0.8 ± 10.4	0.45	0.76
Whole meal Slices (Count/wk)	0.6 ± 2.2	0.5 ± 1.7	0.19	0.38
Milky drinks (Count/wk)	2.7 ± 5.7	2.9 ± 6.4	0.44	0.67
Milk with cereal (Portion/wk)	0.2 ± 1.6	0.1 ± 0.6	0.15	0.95
Cheese (Triangle/wk)	2.5 ± 11.3	2.5 ± 8.8	0.20	0.48
Milk pudding (Portion/wk)	0.4 ± 3.7	0.2 ± 1.5	0.001	0.16
Cottage Cheese (Portion/wk)	1.5 ± 6.0	1.4 ± 6.5	0.45	0.16
Sardines pilchards (Portion/wk)	0.9 ± 7.9	0.5 ± 6.0	0.002	0.05
Fish (Portion/wk)	1.1 ± 7.9	0.7 ± 4.7	0.04	0.87
Eggs (Eggs/wk)	1.2 ± 3.7	0.8 ± 2.4	<0.001	0.60
Cheese omelet (Count/wk)	0.4 ± 6.0	0.5 ± 6.2	0.16	0.83
Porridge (Portion/wk)	0.2 ± 2.1	0.2 ± 0.6	0.87	0.98
Muesli (Portion/wk)	0.9 ± 5.3	0.8 ± 5.6	0.001	0.97
Green vegetable (Portion/wk)	3.5 ± 5.9	3.9 ± 6.8	0.08	0.20

Note: Data presented as N (%) for frequencies; mean ± SD for normal variables and median (quartile 1–quartile 3) for non-normal variables; $p < 0.05$ considered significant. * indicates *p*-values adjusted for age, BMI and gender.

After adjusting for age and BMI, the osteoporosis group appeared to have significantly lower systolic blood pressure (p = 0.05), diastolic blood pressure (p = 0.005), glucose level (p = 0.01) and triglycerides (p = 0.001) than the non-osteoporosis group. However, the osteoporosis group have significantly higher 25(OH)D concentrations than the non-osteoporosis group ($p < 0.001$).

Figure 2 shows the overall mean Ca intake which was 445.1 mg/day, which is far below the recommended dietary intake of 1300 mg/day. Mean Ca intake among women was 464.0 mg/day while in men it was 375.1 mg/day. No significant difference in Ca intake was observed between men and women.

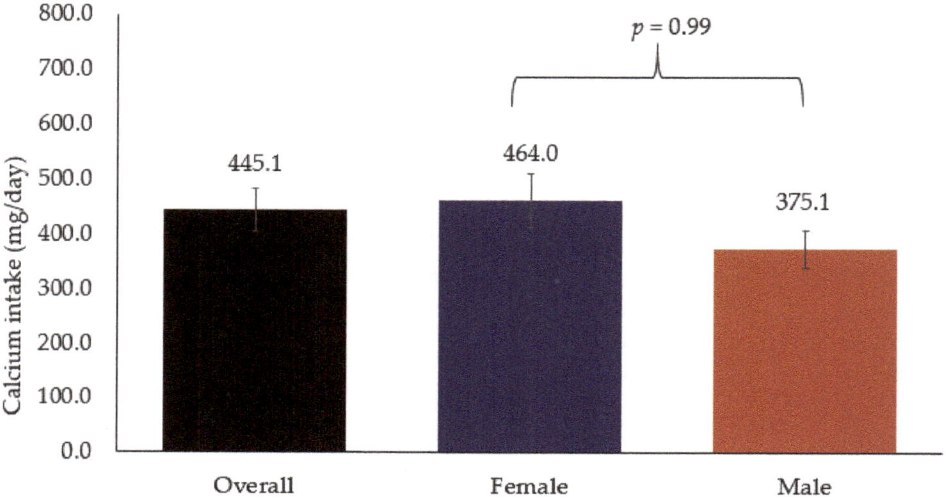

Figure 2. Mean Ca intake among Arab adults.

Table 2 shows the Ca intake from food groups according to osteoporosis status. The Ca intake from tea with milk was significantly higher in the non-osteoporosis group than the osteoporosis groups with odds ratio (OR) of 0.7 (0.6–0.8) ($p < 0.001$). Similarly, the Ca intake of coffee with milk was significantly higher in the non-osteoporosis group than the osteoporosis group with an OR 0.6 (0.5–0.8) ($p < 0.001$). Furthermore, Ca intake from fish and eggs was also significantly higher in the non-osteoporosis group than the osteoporosis group with an OR 0.8 (0.7–0.9) ($p < 0.001$). Ca intake from porridge and oats was also significantly higher in the non-osteoporosis group than osteoporosis group with OR 0.8 (0.8–0.9) (p = 0.005). On the other hand, Ca intake from biscuits, cake and bread slices were significantly higher in the osteoporosis group than the non-osteoporosis group ($p < 0.001$). After adjusting for age and BMI, only consumption of tea with milk, biscuits, cake and bread slices and fish and eggs remained significantly different between groups, while consumption of coffee with milk showed borderline significance (p = 0.05).

Table 3 shows the odds of osteoporosis according to Ca intake from different food groups obtained from multivariate logistic regression analysis. All these variables were log transformed and added simultaneously to the logistic model adjusting for one another. Results showed that consuming tea (p = 0.005), coffee with milk (p = 0.007) and fish and eggs (p = 0.002) were significantly associated with lower risk of osteoporosis. After adjustment with age and BMI, consumption of biscuits, cake and bread slices remained significantly associated with higher risk of osteoporosis (p = 0.02).

Table 2. Ca intake from different food groups according to Osteoporosis status.

Parameters	Non-Osteoporosis		Osteoporosis		Univariate	Adjusted *	*p*-Value	*p*-Value *
	Mean ± SD	Median (Q1–Q3)	Mean ± SD	Median (Q1–Q3)	OR (95%CI)	OR (95%CI)		
Tea with Milk	33.6 ± 117.2	0.0	26.4 ± 132.7	0.0	0.7 (0.6–0.8)	0.8 (0.6–0.9)	<0.001	<0.001
Coffee with Milk	23.9 ± 144.1	0.0	8.5 ± 50.7	0.0	0.6 (0.5–0.8)	0.8 (0.6–1.0)	<0.001	0.05
Biscuits, cake, bread slices	479.5 ± 2920.0	240.0 (110.0–350.0)	432.3 ± 2618.1	260.0 (150.0–360.0)	1.1 (0.9–1.3)	1.2 (1.0–1.5)	0.30	0.02
Milk and milk products	1627.9 ± 4794.6	1000.0 (335.0–2000.0)	1641.3 ± 4389.7	975.0 (320.0–1850.0)	1.0 (0.9–1.1)	1.0 (0.9–1.1)	0.78	0.61
Green vegetables	140.8 ± 235.5	120.0 (40.0–200.0)	154.5 ± 272.4	120.0 (40.0–200.0)	1.1 (1.0–1.3)	1.1 (1.0–1.3)	0.13	0.15
Fish and eggs	601.4 ± 4720.7	50.0 (0.0–222.0)	457.5 ± 5244.5	37.0 (0.0–100.0)	0.8 (0.7–0.9)	0.8 (0.8–0.9)	<0.001	0.001
Porridge and oats	96.4 ± 553.1	0.0 (0.0–100.0)	80.8 ± 561.9	0.0 (0.0–37.0)	0.8 (0.8–0.9)	1.0 (0.8–1.1)	0.005	0.37

Note: Data presented as ca intake from different food groups (mg/wk); OR obtained from logistic linear regression; * indicates *p*-value adjusted for age and BMI.

Table 3. Odds of Osteoporosis according to Ca intake from food groups.

	Model 1		Model 2	
	OR (95%CI)	*p*-Value	OR (95%CI)	*p*-Value
Tea with Milk	0.8 (0.7–0.9)	0.005	0.8 (0.7–1.0)	0.02
Coffee with Milk	0.7 (0.6–0.9)	0.007	0.8 (0.6–1.1)	0.18
Biscuits, Cakes, Pancakes, Bread	1.1 (0.9–1.4)	0.22	1.3 (1.0–1.5)	0.02
Milk and Milk products	1.0 (0.9–1.1)	0.83	0.9 (0.8–1.1)	0.30
Green vegetables	1.2 (1.0–1.4)	0.08	1.1 (0.9–1.3)	0.32
Fish and Eggs	0.8 (0.8–0.9)	0.002	0.9 (0.8–1.0)	0.01
Porridge, Muesli and Oats	0.9 (0.8–1.0)	0.18	1.0 (0.9–1.1)	0.94

Note: Data presented as OR (95%CI); Model 1 include results of multivariate logistic regression; Model 2 include results adjusted for age and BMI; $p < 0.05$ considered significant.

4. Discussion

The present study aimed to determine the association between dietary Ca intake and osteoporosis risk in Saudi adults with known risk factors for osteoporosis. Results showed that the overall mean Ca intake, which was 445.1 mg/day, was substantially below the recommended dietary intake of 1300 mg/day. Mean Ca intake of women was 464.0 mg/day, while in men it was 375.1 mg/day. Dietary Ca intake from tea with milk, fish and eggs were significantly associated with lower risk of osteoporosis while biscuits, cake and bread slices showed a significant association with higher risk of osteoporosis independent of age and BMI.

Ca is an essential divalent cation that is mainly present outside cells and plays a major role in a multitude of vital functions such as muscle contraction, neuronal excitability, neurotransmitter release and blood coagulation. Additionally, the Ca reservoir significantly contributes to the mineral matrix of bone, an indication that the reservoir can influence bone strength [24] and serve as a buffer when required [25]. It is well-established that consuming enough Ca during the period of growth is the most significant nutritional factor for attaining optimal peak bone mass [25–27]. A 2017 systematic review found that the average Ca intake in national diets ranged from 175 to 1233 mg/day, as reported in 78 studies from 74 countries [28]. Similar to the present study, the average dietary Ca intake in many Asian countries is less than 500 mg/day, while countries in Africa and South America generally ranged from 400 to 700 mg/day [28].

Additionally, a more recent systematic review was conducted on the diets of individuals aged 25 and older in 195 countries, which used national or sub-national representative nutrition surveys and multiple dietary data sources from the Global Health Data Exchange [29]. According to this review, the global average daily Ca intake is approximately 400 mg/day, with lower intakes observed in Sub-Saharan African countries and Southeast Asia (around 200 mg/day) as compared to high-income countries (around 600–800 mg/day) [29]. In this study, the average Ca intake in adult males was 375.1 mg/d, while for females it was 464.0 mg/d. Unfortunately, inadequate Ca intake and suboptimal levels of vitamin D are prevalent among the elderly population of SA, with the latter improving over time across all demographic groups due to successful public health campaigns [30]. It is expected that this improvement in vitamin D levels may also bring a general improvement in Ca intake through continuous public health campaigns.

This study suggests that Ca intake from tea with milk is significantly associated with decreased risk of osteoporosis. Some studies reported that frequent tea and coffee intake had little influence on BMD and were not associated with fracture risk among postmenopausal women with osteoporosis [31,32]. In a meta-analysis which included 17 studies [14], the odds of osteoporosis for the highest versus the lowest categories of tea consumption was 0.62 (95%CI, 0.46–0.83). Furthermore, the risk-lowering effect of tea consumption on osteoporosis was consistent in males and females, different study periods, study designs and multiple geographic locations [14]. This beneficial effect of tea consumption with milk may be due to the presence of polyphenols that are responsible for improving BMD by reducing oxidative stress and inflammation [33]. Additionally, polyphenols are known to be sequestered into physiological bone, which may help protect it from degradation [14,33]. While more research is needed to understand the exact mechanisms by which polyphenols may help protect bone health, it is suggested that consuming polyphenol-rich foods, such as tea, may help protect against bone loss.

The present study also found the protective effects of fish and eggs consumption on osteoporosis risk. Both fish and eggs are known to be great sources of protein which help build and maintain strong bones and muscles. Pujia et al. [17] also reported a positive association between whole egg consumption and bone health. Furthermore, numerous studies have shown that fish consumption is linked to decreased osteoporosis risk, increased BMD and decreased fracture risk for both men and women—more so in Asian populations [34–37] than Caucasians, as observed in the Cardiovascular Health Study [38].

In the present study, Ca intake from cake, biscuits and bread contributed to increased osteoporosis risk. These results are consistent with other observations which found that high glycemic index (GI) foods had negative consequences for bone health, including decreased osteoblast activity, increased bone resorption, oxidative stress and inflammation, as well as the induction of acidosis [39,40]. In contrast, complex carbohydrates such as those found in vegetables, legumes and whole grains can increase Ca absorption and neutralize metabolic acid loads in the body [9]. Inulin-type fructans are a form of non-digestible carbohydrate found in various vegetables and fruits, which cannot be broken down by mammalian enzymes [9]. Inulin-type fructans alter intestinal Ca absorption, enhance Ca solubility in the lumen and amplify bioavailability, eventually leading to improved passive transport of into the body [9,41]. While the present study did not observe an association between vegetable and fruit consumption on osteoporosis risk, it should not supersede previous findings on the beneficial effects of vegetables and fruits consumption on bone health [41–43].

Large populations across the globe consider milk and dairy products, including cheese, butter, cream and yogurt, to be major sources of Ca, which can increase BMD and potentially reduce the risk of fractures, particularly in postmenopausal women [15,40,44]. A recent 2021 systematic review of 8 case-control studies, 10 cross-sectional studies, 17 prospective cohort studies and 14 randomized clinical trials (RCT) reported that dairy consumption had a moderate effect on bone health in older adults aged 50 years or more [44]. All case-control studies showed reduced hip fracture risk with the exception of one study; the included cross-sectional studies showed favorable impact of milk intake on BMD; and RCTs showed a beneficial impact of fortified dairy products [44]. In the same review, however, there is ambiguity as to the life period in which milk intake is most advantageous [44].

It is important to note that the relationship between dairy intake and bone health is also influenced by other factors such as exercise, genetics and overall nutrition [45,46]. For instance, hip fracture risk in the elderly was found to be higher in those who self-reported consuming milk and dairy products, specifically at the age of 20 years, suggesting a potential negative impact of high dairy intake [46]. In Feskanich et al.'s investigation, Ca supplements may help prevent osteoporotic fractures in certain bone areas, but regularly consuming high amounts of Ca-rich foods during middle age is unlikely to confer any protective benefits against hip or forearm fractures, at least for white middle-aged women in the US [45]. The mentioned study was followed-up after 6 years and involved analysis of 72,337 postmenopausal women where a total of 603 hip fractures occurred due to low or moderate trauma during the period from 1980 to 1998 [47]. The risk of hip fracture was not linked to either milk intake or total Ca consumed, and instead there was a lower risk of osteoporotic hip fractures in postmenopausal women with sufficient intake of vitamin D [47]. The results of our study also showed no association between Ca intake from dairy products and risk of osteoporosis. These inconsistencies in the research regarding a potential link between milk consumption and bone health can be attributed, in part, to the biases and inaccuracies inherent in dietary data collection [48].

The present study has certain limitations which should be taken into consideration when interpreting findings. First, the results are based on observational study; thus, causality cannot be inferred. Second, Ca intake from supplements and medications in general were not recorded and this could have explained why the osteoporosis group appeared to have a better metabolic profile (e.g., lower BP and glucose, higher 25(OH)D) than the non-osteoporosis group. Finally, the findings of the present study are most relevant to Arab adults already at risk of osteoporosis. Similar studies using other populations (e.g., pregnant, pre-menopausal) in the same ethnic group may be able to provide additional information on the over-all dietary deficiencies present in the modern-day Saudi diet.

5. Conclusions

In conclusion, extremely low dietary Ca intake was observed among Saudi adults at risk of osteoporosis. Ca intake from tea with milk, fish and eggs decrease, while Ca intake from cakes, biscuits and bread increase, the risk of osteoporosis. These findings suggest an urgent need to promote Ca supplementation in this high-risk population since diet alone is clearly not enough to meet daily Ca needs. A balanced diet that meets the recommended intake of Ca, vitamin D and omega-3 fatty acids, as well as limiting foods high in saturated fats and/or glycemic index, may be effective in reducing osteoporosis risk in the Saudi adult population.

Author Contributions: Conceptualization S.S.; methodology, S.S., S.D.H., A.M.A. and N.A.; formal analysis, S.D.H. writing—original draft preparation, S.D.H.; writing—review and editing, S.S. and N.M.A.-D.; supervision, N.M.A.-D.; funding acquisition, N.M.A.-D. All authors have read and agreed to the published version of the manuscript.

Funding: The study was funded by the Deputyship for Research and Innovation, Ministry of Education in Saudi Arabia for funding this research (IFKSURC-1-1604).

Institutional Review Board Statement: The study was conducted according to the guidelines of the Declaration of Helsinki and approved by the Ethics Committee of the College of Medicine, King Saud University, Riyadh, Kingdom of Saudi Arabia (approval number H-01-R-012).

Informed Consent Statement: Written Informed consent was obtained from each participant involved in the study.

Data Availability Statement: The data presented in this study are available on reasonable request from the corresponding author.

Acknowledgments: The authors extend their appreciation to the Deputyship for Research and Innovation, Ministry of Education in Saudi Arabia, for funding this research (IFKSURC-1-1604).

Conflicts of Interest: The authors declare no conflict of interest.

References

1. Al-Daghri, N.M.; Sabico, S.; Al-Saleh, Y.; Sulimani, R.; Aljohani, N.J.; Sheshah, E.; Alodhayani, A.; Harvey, N.C.; Liu, E Lorentzon, M.; et al. The application of FRAX in Saudi Arabia. *Arch. Osteoporos.* **2021**, *16*, 166. [CrossRef]
2. Kanis, J.A.; McCloskey, E.V.; Johansson, H.; Oden, A.; Melton, L.J., 3rd; Khaltaev, N. A reference standard for the description o osteoporosis. *Bone* **2008**, *42*, 467–475. [CrossRef]
3. Sadat-Ali, M.; AlZamami, J.F.; AlNaimi, S.N.; Al-Noaimi, D.A.; AlDakheel, D.A.; AlSayed, H.N.; Al-Turki, H.A.; AlOmran, A.S Osteoporosis: Is the prevalence increasing in Saudi Arabia. *Ann. Afr. Med.* **2022**, *21*, 54–57. [CrossRef] [PubMed]
4. Al Hamam, N.M.; Al-Moaibed, G.F.; Alfayez, E.H.; Alfayez, E.H.; Al-Mubaddil, M.S.; Alramadhan, N.A. Prevalence and ris factors for osteoporotic fracture among adults with comorbidities in Al-Ahsaa, Saudi Arabia. *J. Fam. Med. Prim. Care* **202** *9*, 877–882. [CrossRef]
5. Rizzoli, R.; Bianchi, M.L.; Garabedian, M.; McKay, H.A.; Moreno, L.A. Maximizing bone mineral mass gain during growth for th prevention of fractures in the adolescents and the elderly. *Bone* **2010**, *46*, 294–305. [CrossRef]
6. Ferrari, S.; Rizzoli, R.; Bonjour, J.P. Genetic aspects of osteoporosis. *Curr. Opin. Rheumatol.* **1999**, *11*, 294–300. [CrossRef] [PubMed
7. Rizzoli, R. Nutrition: Its role in bone health. *Best Pract. Res. Clin. Endocrinol. Metab.* **2008**, *22*, 813–829. [CrossRef] [PubMed]
8. Rizzoli, R. Dairy products, yogurts, and bone health. *Am. J. Clin. Nutr.* **2014**, *99*, 1256S–1262S. [CrossRef]
9. Lorincz, C.; Manske, S.L.; Zernicke, R. Bone health: Part 1, nutrition. *Sport. Health* **2009**, *1*, 253–260. [CrossRef]
10. Munoz-Garach, A.; Garcia-Fontana, B.; Munoz-Torres, M. Nutrients and Dietary Patterns Related to Osteoporosis. *Nutrients* **202** *12*, 1986. [CrossRef]
11. Malmir, H.; Saneei, P.; Larijani, B.; Esmaillzadeh, A. Adherence to Mediterranean diet in relation to bone mineral density and ris of fracture: A systematic review and meta-analysis of observational studies. *Eur. J. Nutr.* **2018**, *57*, 2147–2160. [CrossRef]
12. Matia-Martin, P.; Torrego-Ellacuria, M.; Larrad-Sainz, A.; Fernandez-Perez, C.; Cuesta-Triana, F.; Rubio-Herrera, M.A. Effects o Milk and Dairy Products on the Prevention of Osteoporosis and Osteoporotic Fractures in Europeans and Non-Hispanic White from North America: A Systematic Review and Updated Meta-Analysis. *Adv. Nutr.* **2019**, *10*, S120–S143. [CrossRef] [PubMed]
13. Fung, T.T.; Feskanich, D. Dietary patterns and risk of hip fractures in postmenopausal women and men over 50 years. *Osteoporo Int.* **2015**, *26*, 1825–1830. [CrossRef] [PubMed]
14. Sun, K.; Wang, L.; Ma, Q.; Cui, Q.; Lv, Q.; Zhang, W.; Li, X. Association between tea consumption and osteoporosis: A meta-analysis. *Medicine* **2017**, *96*, e9034. [CrossRef]
15. Pujia, R.; Ferro, Y.; Maurotti, S.; Mare, R.; Arturi, F.; Montalcini, T.; Pujia, A.; Mazza, E. Relationship between Osteoporosis Multiple Fractures, and Egg Intake in Healthy Elderly. *J. Midlife Health* **2021**, *12*, 287–293. [CrossRef] [PubMed]
16. Calderon-Garcia, J.F.; Moran, J.M.; Roncero-Martin, R.; Rey-Sanchez, P.; Rodriguez-Velasco, F.J.; Pedrera-Zamorano, J.D. Dietar habits, nutrients and bone mass in Spanish premenopausal women: The contribution of fish to better bone health. *Nutrients* **201** *5*, 10–22. [CrossRef]
17. Sabico, S.; Wani, K.; Grant, W.B.; Al-Daghri, N.M. Improved HDL Cholesterol through Vitamin D Status Correction Substantiall Lowers 10-Year Atherosclerotic Cardiovascular Disease Risk Score in Vitamin D-Deficient Arab Adults. *Nutrients* **2023**, *15*, 55 [CrossRef]
18. Alissa, E.M.; Qadi, S.G.; Alhujaili, N.A.; Alshehri, A.M.; Ferns, G.A. Effect of diet and lifestyle factors on bone health i postmenopausal women. *J. Bone Miner. Metab.* **2011**, *29*, 725–735. [CrossRef]
19. Al-Daghri, N.M.; Yakout, S.; Sabico, S.; Wani, K.; Hussain, S.D.; Aljohani, N.; Uday, S.; Hogler, W. Establishing the Prevalence c Osteomalacia in Arab Adolescents Using Biochemical Markers of Bone Health. *Nutrients* **2022**, *14*, 5354. [CrossRef]
20. Wani, K.; Yakout, S.M.; Ansari, M.G.A.; Sabico, S.; Hussain, S.D.; Alokail, M.S.; Sheshah, E.; Aljohani, N.J.; Al-Saleh, Y Reginster, J.Y.; et al. Metabolic Syndrome in Arab Adults with Low Bone Mineral Density. *Nutrients* **2019**, *11*, 1405. [CrossRef]
21. Al-Saleh, Y.; Al-Daghri, N.M.; Sabico, S.; Alessa, T.; Al Emadi, S.; Alawadi, F.; Al Qasaabi, S.; Alfutaisi, A.; Al Izzi, M Mukhaimer, J.; et al. Diagnosis and management of osteoporosis in postmenopausal women in Gulf Cooperation Council (GCC countries: Consensus statement of the GCC countries' osteoporosis societies under the auspices of the European Society fo Clinical and Economic Aspects of Osteoporosis and Osteoarthritis (ESCEO). *Arch. Osteoporos.* **2020**, *15*, 109. [CrossRef]
22. Al-Saleh, Y.; Sulimani, R.; Sabico, S.; Alshahrani, F.M.; Fouda, M.A.; Almohaya, M.; Alaidarous, S.B.; Alkhawashki, H.M Alshaker, M.; Alrayes, H.; et al. Diagnosis and management of osteoporosis in Saudi Arabia: 2023 key updates from the Saud Osteoporosis Society. *Arch. Osteoporos.* **2023**, *18*, 75. [CrossRef] [PubMed]
23. Macdonald, H.M.; Garland, A.; Burr, J.; Strachan, A.; Wood, A.D.; Jamil, N.A.; McLernon, D.; Black, A.J. Validation of a shor questionnaire for estimating dietary calcium intakes. *Osteoporos. Int.* **2014**, *25*, 1765–1773. [CrossRef]
24. Vannucci, L.; Fossi, C.; Quattrini, S.; Guasti, L.; Pampaloni, B.; Gronchi, G.; Giusti, F.; Romagnoli, C.; Cianferotti, L Marcucci, G.; et al. Calcium Intake in Bone Health: A Focus on Calcium-Rich Mineral Waters. *Nutrients* **2018**, *10*, 1930. [CrossRef
25. Caroli, A.; Poli, A.; Ricotta, D.; Banfi, G.; Cocchi, D. Invited review: Dairy intake and bone health: A viewpoint from the state o the art. *J. Dairy Sci.* **2011**, *94*, 5249–5262. [CrossRef] [PubMed]

26. Anderson, J.J.; Tylavsky, F.A.; Halioua, L.; Metz, J.A. Determinants of peak bone mass in young adult women: A review. *Osteoporos. Int.* **1993**, *3* (Suppl. S1), 32–36. [CrossRef] [PubMed]
27. Sowers, M.R.; Galuska, D.A. Epidemiology of bone mass in premenopausal women. *Epidemiol. Rev.* **1993**, *15*, 374–398. [CrossRef]
28. Balk, E.M.; Adam, G.P.; Langberg, V.N.; Earley, A.; Clark, P.; Ebeling, P.R.; Mithal, A.; Rizzoli, R.; Zerbini, C.A.F.; Pierroz, D.D.; et al. Global dietary calcium intake among adults: A systematic review. *Osteoporos. Int.* **2017**, *28*, 3315–3324. [CrossRef]
29. Collaborators, G.B.D.D. Health effects of dietary risks in 195 countries, 1990–2017: A systematic analysis for the Global Burden of Disease Study 2017. *Lancet* **2019**, *393*, 1958–1972. [CrossRef]
30. Al-Daghri, N.M.; Hussain, S.D.; Ansari, M.G.A.; Khattak, M.N.K.; Aljohani, N.; Al-Saleh, Y.; Al-Harbi, M.Y.; Sabico, S.; Alokail, M.S. Decreasing prevalence of vitamin D deficiency in the central region of Saudi Arabia (2008–2017). *J. Steroid Biochem. Mol. Biol.* **2021**, *212*, 105920. [CrossRef]
31. Chen, Z.; Pettinger, M.B.; Ritenbaugh, C.; LaCroix, A.Z.; Robbins, J.; Caan, B.J.; Barad, D.H.; Hakim, I.A. Habitual tea consumption and risk of osteoporosis: A prospective study in the women's health initiative observational cohort. *Am. J. Epidemiol.* **2003**, *158*, 772–781. [CrossRef] [PubMed]
32. Barrett-Connor, E.; Chang, J.C.; Edelstein, S.L. Coffee-associated osteoporosis offset by daily milk consumption. The Rancho Bernardo Study. *JAMA* **1994**, *271*, 280–283. [CrossRef] [PubMed]
33. Park, Y.H.; Han, D.W.; Suh, H.; Ryu, G.H.; Hyon, S.H.; Cho, B.K.; Park, J.C. Protective effects of green tea polyphenol against reactive oxygen species-induced oxidative stress in cultured rat calvarial osteoblast. *Cell Biol. Toxicol.* **2003**, *19*, 325–337. [CrossRef]
34. Farina, E.K.; Kiel, D.P.; Roubenoff, R.; Schaefer, E.J.; Cupples, L.A.; Tucker, K.L. Protective effects of fish intake and interactive effects of long-chain polyunsaturated fatty acid intakes on hip bone mineral density in older adults: The Framingham Osteoporosis Study. *Am. J. Clin. Nutr.* **2011**, *93*, 1142–1151. [CrossRef]
35. Nawata, K.; Yamauchi, M.; Takaoka, S.; Yamaguchi, T.; Sugimoto, T. Association of *n*-3 polyunsaturated fatty acid intake with bone mineral density in postmenopausal women. *Calcif. Tissue Int.* **2013**, *93*, 147–154. [CrossRef] [PubMed]
36. Fan, F.; Xue, W.Q.; Wu, B.H.; He, M.G.; Xie, H.L.; Ouyang, W.F.; Tu, S.L.; Chen, Y.M. Higher fish intake is associated with a lower risk of hip fractures in Chinese men and women: A matched case-control study. *PLoS ONE* **2013**, *8*, e56849. [CrossRef] [PubMed]
37. Chen, Y.M.; Ho, S.C.; Lam, S.S. Higher sea fish intake is associated with greater bone mass and lower osteoporosis risk in postmenopausal Chinese women. *Osteoporos. Int.* **2010**, *21*, 939–946. [CrossRef]
38. Virtanen, J.K.; Mozaffarian, D.; Cauley, J.A.; Mukamal, K.J.; Robbins, J.; Siscovick, D.S. Fish consumption, bone mineral density, and risk of hip fracture among older adults: The cardiovascular health study. *J. Bone Miner. Res.* **2010**, *25*, 1972–1979. [CrossRef]
39. Nouri, M.; Mahmoodi, M.; Shateri, Z.; Ghadiri, M.; Rajabzadeh-Dehkordi, M.; Vali, M.; Gargari, B.P. How do carbohydrate quality indices influence on bone mass density in postmenopausal women? A case-control study. *BMC Womens Health* **2023**, *23*, 42. [CrossRef]
40. Garcia-Gavilan, J.F.; Bullo, M.; Camacho-Barcia, L.; Rosique-Esteban, N.; Hernandez-Alonso, P.; Basora, J.; Martinez-Gonzalez, M.A.; Estruch, R.; Fito, M.; Salas-Salvado, J. Higher dietary glycemic index and glycemic load values increase the risk of osteoporotic fracture in the PREvencion con DIeta MEDiterranea (PREDIMED)-Reus trial. *Am. J. Clin. Nutr.* **2018**, *107*, 1035–1042. [CrossRef]
41. Coxam, V. Current data with inulin-type fructans and calcium, targeting bone health in adults. *J. Nutr.* **2007**, *137*, 2527S–2533S. [CrossRef] [PubMed]
42. Raschka, L.; Daniel, H. Mechanisms underlying the effects of inulin-type fructans on calcium absorption in the large intestine of rats. *Bone* **2005**, *37*, 728–735. [CrossRef]
43. Qiu, R.; Cao, W.T.; Tian, H.Y.; He, J.; Chen, G.D.; Chen, Y.M. Greater Intake of Fruit and Vegetables Is Associated with Greater Bone Mineral Density and Lower Osteoporosis Risk in Middle-Aged and Elderly Adults. *PLoS ONE* **2017**, *12*, e0168906. [CrossRef] [PubMed]
44. Wallace, T.C.; Bailey, R.L.; Lappe, J.; O'Brien, K.O.; Wang, D.D.; Sahni, S.; Weaver, C.M. Dairy intake and bone health across the lifespan: A systematic review and expert narrative. *Crit. Rev. Food Sci. Nutr.* **2021**, *61*, 3661–3707. [CrossRef] [PubMed]
45. Feskanich, D.; Willett, W.C.; Stampfer, M.J.; Colditz, G.A. Milk, dietary calcium, and bone fractures in women: A 12-year prospective study. *Am. J. Public Health* **1997**, *87*, 992–997. [CrossRef] [PubMed]
46. Cumming, R.G.; Klineberg, R.J. Case-control study of risk factors for hip fractures in the elderly. *Am. J. Epidemiol.* **1994**, *139*, 493–503. [CrossRef] [PubMed]
47. Feskanich, D.; Willett, W.C.; Colditz, G.A. Calcium, vitamin D, milk consumption, and hip fractures: A prospective study among postmenopausal women. *Am. J. Clin. Nutr.* **2003**, *77*, 504–511. [CrossRef] [PubMed]
48. Nel, J.H.; Steyn, N.P.; Senekal, M. Illustration of the Importance of Adjustment for within- and between-Person Variability in Dietary Intake Surveys for Assessment of Population Risk of Micronutrient Deficiency/Excess Using an Example Data Set. *Nutrients* **2022**, *14*, 285. [CrossRef]

Disclaimer/Publisher's Note: The statements, opinions and data contained in all publications are solely those of the individual author(s) and contributor(s) and not of MDPI and/or the editor(s). MDPI and/or the editor(s) disclaim responsibility for any injury to people or property resulting from any ideas, methods, instructions or products referred to in the content.

nutrients

Article

Vitamin D Status among Patients Admitted to a Geriatric Ward—Are Recommendations for Preventing Its Deficiency Effective Enough?

Maksymilian Adam Lech [1], Marcin Warpechowski [1], Aleksandra Wojszel [1], Justyna Rentflejsz [2,3,4], Marta Świętek [4] and Zyta Beata Wojszel [3,4,*]

1 Interdisciplinary Student Scientific Society, Department of Geriatrics, Medical University of Bialystok, 15-471 Bialystok, Poland; maksymilian.lech1@gmail.com (M.A.L.); warpechowski.marcin@gmail.com (M.W.); aleksandrawojszel@gmail.com (A.W.)
2 Doctoral School, Medical University of Bialystok, 15-089 Bialystok, Poland; justyna.rentflejsz@sd.umb.edu.pl
3 Department of Geriatrics, Medical University of Bialystok, 15-471 Bialystok, Poland
4 Department of Geriatrics and Internal Medicine, Hospital of the Ministry of Interior in Bialystok, 15-471 Bialystok, Poland; kmar@vp.pl
* Correspondence: wojszel@umb.edu.pl; Tel.: +48-47-710-4242

Abstract: Despite a decade of available recommendations aimed at preventing vitamin D (VD) deficiency in Poland, the efficacy of these measures among community-dwelling older individuals remains inconsistent. The PolSenior2 study provided valuable insights into VD status among community-dwelling older individuals in Poland. However, it is important to note that this research did not include the elderly living in care institutions. Therefore, our study concentrates on evaluating VD status in older patients admitted to a geriatrics ward to indirectly assess their adherence to existing recommendations and preventive actions in this particular setting (whether they translate into health-promoting behaviors (i.e., taking vitamin D supplements) and whether the recommended, optimal 25(OH)D concentration values are achieved). This approach offers a comprehensive understanding of VD status in a previously understudied population. We aimed to evaluate VD status in patients aged 70 and above within the geriatrics ward, exploring its association with age, sex, BMI, and the use of VD supplements. The study involved the measurement of serum VD concentration in 240 individuals. Of these participants, 177 (73.8%) were women, and 193 (80.4%) were over 75 years old. The median 25(OH)D concentration was found to be 22.95 (IQR, 13.7–33.0) ng/mL. Notably, profound deficiency (<10 ng/mL) was noted in 15% of the participants, while 67.5% exhibited VD inadequacy (<30 ng/mL). It is worth mentioning that only 18.3% of individuals took VD supplements preadmission. Insufficiency was more prevalent in nonsupplemented individuals (70.9% vs. 52.3%, $p = 0.02$) and those with a BMI $\geq$30 kg/m^2 (76.2% vs. 59.2%, $p = 0.007$). The logistic regression model demonstrated that obese patients had over two times higher odds of VD inadequacy (OR = 2.21, $p = 0.0074$), as did nonsupplemented individuals (OR = 2.23, $p = 0.0187$). The high prevalence of VD deficiency and inadequacy in geriatric ward admissions emphasizes the urgent need for targeted interventions and enhanced education for older adults, caregivers, and physicians to improve adherence to preventive supplementation practices.

Keywords: vitamin D; vitamin D deficiency; vitamin D inadequacy; aged; frail older adults; dietary supplements; supplementation; odds ratio; logistic regression; recommendations

1. Introduction

Vitamin D (VD) constitutes a group of fat-soluble steroid chemicals that play an essential role in various physiological processes. It influences and maintains the condition of the skeletal system as its primary function is related to calcium–phosphate metabolism. Deficiency of VD in older adults is associated with osteoporosis, osteomalacia, myalgia, and sarcopenia. Studies focused on the holistic understanding of the effect of VD on the human body suggest that maintaining an optimal VD level, coupled with physical activity and proper nutrition, may mitigate the risk of musculoskeletal system diseases. This, in turn, positively impacts the overall quality of life for older individuals [1]. A low concentration of 25(OH)D in the serum (an indicator of VD resources in the body) is associated with a higher incidence of cardiovascular diseases, compromised immune system functioning, depression, and neurodegenerative disorders such as Alzheimer's disease [1–5]. In addition, some studies indicate that maintaining an appropriate VD level reduces the risk of developing colon, breast, ovarian, and prostate cancer [6]. However, it is crucial to note that while a meta-analysis of randomized trials showed that it had no role or effect on cancer incidence, it demonstrated a reduction in cancer-related mortality once cancer was diagnosed [7]. Despite these promising observations regarding various cancers, there is an evident need for well-designed research in this area [8]. In light of the current knowledge, it is clear that VD plays a significant role in the body, and efforts should be made to maintain its appropriate level in the serum.

In human plasma, VD is present in the forms of cholecalciferol (vitamin D3) or ergocalciferol (vitamin D2). Cholecalciferol is formed in the skin from 7-dehydrocholesterol under the influence of UV rays (particularly solar ultraviolet B radiation (UV-B)), and ergocalciferol is obtained as a component of plant-based foods [7]. Both variants exhibit similar biological activities. Within the liver, both cholecalciferol and ergocalciferol undergo 25-hydroxylation to form 25-hydroxycholecalciferol and 25-hydroxyergocalciferol, respectively. Subsequently, in the kidney, 1-alpha-hydroxylation occurs to produce 1,25-dihydroxyergocalciferol and the more metabolically active 1,25-dihydroxycholecalciferol (calcitriol). The enzyme 24-hydroxylase further converts 25(OH)D to 24,25(OH)2D, an inactive metabolite (Figure 1) [9].

VD participates in the regulation of calcium and phosphate metabolism, influencing bone metabolism, promoting bone mineralization by osteoblasts, and increasing the absorption of calcium and phosphates in the intestine. Its role in the pathogenesis of several diseases also results from its so-called "nonclassical," pleiotropic effects. The presence of VD 1-hydroxylase in the cells of many tissues (known as peripheral hydroxylases) and local, peripheral production of 1,25(OH)2D have been demonstrated [9]. VD exerts its effects by activating the membrane receptor (MARRS, membrane-associated, rapid response steroid-binding) and VD receptors (VDR) present in the nuclei of most cells. These transcription factors regulate the expression of over 2000 genes responsible for cell differentiation, proliferation, and apoptosis [10]. A significant portion of the VD in the human body is produced in the skin following exposure to solar ultraviolet B radiation (UV-B) [11]. In the absence of sufficient biosynthesis in the skin, which is common among the older population, dietary intake becomes the most critical source of VD.

The primary sources of VD in the diet are the few products that naturally contain VD (e.g., fatty fish, beef, egg, egg yolk, veal, mushrooms, liver), fortified foods (cereal and dairy products), and supplements [12]. Notably, a deficiency in VD has been associated with an increased risk of cancer development [13]. Recognizing the importance of VD and a healthy diet, the World Cancer Research Fund recommends them as key elements for cancer prevention. Following these guidelines has shown a significant reduction in the risk of developing colorectal cancer, as indicated by the subjects of a recent study conducted by Serrano and co-workers [14]. According to the current recommendations, VD supplementation based on cholecalciferol for older adults is recommended throughout the year. For younger seniors (65–75 years), due to decreased efficacy of the skin synthesis, a dose of 1000–2000 IU/day (25–50 μg/day) is recommended; for older seniors (>75–89 years) and the oldest individuals (90 years and older), due to decreased efficacy of the skin synthesis, potential malabsorption and altered metabolism of VD, a higher dose of 2000–4000 IU/day (50–100 μg/day) is recommended [15,16]. When determining the quantity individually, one should account for factors such as age, body weight, VD intake from food, comorbidities, and medications taken [15]. An essential aspect when determining the dosage is obesity. Recommendations for supplementation in individuals diagnosed with obesity indicate the need to double the daily VD intake amount compared to those without obesity. Therefore, obese people over 75 should take 4000–8000 IU/day (100–200 μg/day). In cases of VD deficiency, doses should be appropriately higher, adjusted to the degree of deficiency (therapeutic doses) [15–17].

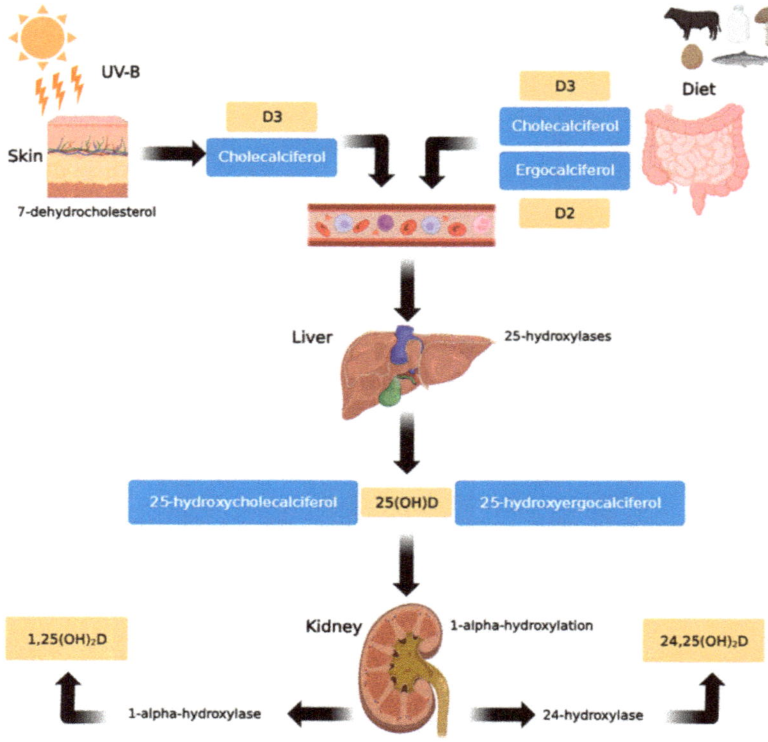

Figure 1. Vitamin D metabolic pathway (created using GIMP 2.10 software). In the skin, exposure to sunlight (UV-B) transforms provitamin D3 (7-dehydrocholesterol) into previtamin D3. Vitamin D3 is produced from previtamin D3 in the basal layers of the epidermis and is absorbed from natural and fortified foods and supplements (D2, D3) in the intestines. It then binds to vitamin D binding protein (DBP) in the bloodstream, before being transported to the liver and hydroxylated there by 25-hydroxylases. In the kidneys, 25-hydroxycholecalciferol [25(OH)D] (calcifediol or calcidiol) is hydroxylated by 1α-hydroxylase, resulting in the formation of the active secosteroid 1,25(OH)2D (calcitriol), which then affects various target tissues. The synthesis of 1,25(OH)2D is stimulated by parathyroid hormone and inhibited by calcium, phosphates, and 1,25(OH)2D [18].

Despite widely available knowledge regarding the importance of VD for human health and published international recommendations aimed at preventing VD deficiency, it remains a global public health challenge affecting specific vulnerable groups, including pregnant women, children with low birth weight, and older adults. The prevalence of VD deficiency remains unknown. It depends on factors such as measurement methodology, criteria for deficiency classification, socio-economic status, and the age of the population under study [19]. Published data indicate that approximately 40% of the European population has a VD concentration (assessed as serum 25(OH)D) not exceeding 20 ng/mL (<50 nmol/l), and 40–80% of the older population struggle with VD deficiency [20]. High percentages of inadequate VD levels among seniors are found regardless of gender, skin color, or place of residence [21]. The nationwide PolSenior 2 study conducted in Poland from 2016 to 2020 showed that VD deficiency (25(OH)D concentration <20 ng/mL) was present in half of the participants over 60. The authors of the study indicated the need to consider VD supplementation when establishing the fundamental canon of public health activities for seniors in Poland [22–24]. Thanks to the PolSenior2 study, we now have up-to-date and relatively comprehensive information on VD status in Polish community-dwelling older people. However, the research did not include the elderly living in care institutions, or those who are hospitalized, more disabled, or diseased. Therefore, our study aims to assess VD concentrations in older patients admitted to a geriatric ward and to evaluate the adherence to existing recommendations and preventive actions in this particular setting (whether they translate into health-promoting behaviors (i.e., taking vitamin D supplements) and whether the recommended, optimal 25 (OH) D concentration values are achieved). This assessment may offer insights into the effectiveness of current recommendations within this specific group, indicating potential areas for additional actions in addressing VD deficiency.

2. Materials and Methods

This study included patients admitted to the Department of Geriatrics at the Hospital of the Ministry of Interior and Administration in Bialystok, Poland, between May 2019 and January 2020. One of the admission criteria, in alignment with the principles of the National Health Fund, was a minimum age of 60 years. However, the specific health issues of individuals referred to geriatrics departments often result in the majority being over 70 years old [25–27].

Each patient underwent a comprehensive geriatric assessment (CGA), which included a detailed medical history, physical examination, and laboratory tests. One of the crucial elements was evaluating the patient's nutritional status, with a focus on the levels of essential vitamins. In the analysis, we considered socio-demographic data (age, gender, place of residence), anthropometric parameters (BMI), and VD concentration. We specifically included patients for whom 25-hydroxyvitamin D (25(OH)D, including 25(OH)D2 and 25(OH)D3) levels were determined, considering it the most reliable indicator of VD status [16]. Total vitamin D/25(OH)D concentration measurements were conducted using the Cobas Pure e402 immunochemical analyzer (manufacturer Roche). The chemiluminescence method, based on an immunochemical reaction using the biotin–streptavidin system, was employed for accurate assessment.

In addition to these measurements, information was gathered on the intake of VD supplements when patients were admitted to the hospital. These data were considered indicative of compliance with recommendations for supplementation and the prevention of VD deficiency. The recommendations in force in Poland at the time of the study stated that older adults should be constantly supplemented with vitamin D throughout the year, in a dose adapted to body weight (higher doses for obese people) and age. Seniors (65–75 Years), due to decreased efficacy of the skin synthesis, should receive a dose of 800–2000 IU/day. The eldest seniors (>75 Years), due to decreased efficacy of the skin synthesis, potential malabsorption, and altered metabolism of vitamin D, should be provided with a dose of 2000–4000 IU/day. The optimal range of vitamin D concentration, which should be strived for (25 (OH)D = 30-50 ng/mL), suboptimal concentration (>20–30 ng/mL), deficiency (>10–20 ng/mL), and severe deficiency (0–10 ng/mL) were also indicated, as well as high concentrations (>50–100 ng/mL) and toxic concentrations (>100 ng/mL) [16].

2.1. Study Parameters

Following the most recent guidelines for the prevention and treatment of VD deficiency in Poland [15–17], patients were stratified according to their 25(OH)D concentration as follows:

- Severe VD deficiency: level in the range of 0–10 ng/mL;
- VD deficiency: >10–20 ng/mL;
- Suboptimal level (insufficiency): >20–30 ng/mL;

- Optimal range: >30–50 ng/mL;
- High level: >50–100 ng/mL;
- Toxic range: >100 ng/mL.

For the current analysis, a concentration of 25(OH)D <30 ng/mL was treated as VD inadequacy because a concentration of VD >30 ng/mL is regarded as the optimal level one should aim for to achieve optimal benefits from VD status.

BMI [kg/m^2] was stratified according to World Health Organization guidelines [28] as follows:

- <18.5—underweight;
- 18.5–24.99—normal weight;
- 25.0–29.99—overweight;
- 30 and over—obesity.

2.2. Statistical Analysis

The data collected for analysis were processed using the IBM SPSS Statistica Version 27 Software suite (SPSS, Chicago, IL, USA) and R version 4.3.1. The Kolmogorov–Smirnov test was used to assess the distribution of variables. Data were presented as mean (M) and standard deviation (SD) for normally distributed continuous variables, median (Me) and interquartile range (IQR) for non-normally distributed ones, and the number of cases and percentages for categorical variables. Proportions were compared using χ^2 tests or Fisher's exact test, as appropriate, while the independent samples Student's *t*-test and the Mann–Whitney U-test were used to compare the distribution of continuous variables. For comparisons among more than two groups, Kruskal–Wallis and post hoc tests were used. Logistic regression (direct method) was used to create a multivariable model to assess the association between VD inadequacy and selected predictors. Odds ratios with a 95% confidence interval (CI) and *p*-value were estimated. Only VD deficiency predictors with a *p*-value lower than 0.1 were included in the model. The cut-off point was set at 0.5, and each model underwent ten-fold cross-validation. Missing values were omitted, and statistics in such cases were calculated for the adequately reduced groups. A two-tailed *p*-value of less than 0.05 was considered statistically significant in all analyses.

2.3. Ethics Approval

The study was approved by the Ethics Committee at the Medical University of Bialystok (no R-I-002/326/2017). All procedures performed adhered to the ethical standards set by the Medical University of Bialystok Ethics Committee in accordance with the Helsinki Declaration and its subsequent amendments. This study adheres to the usual practice protocols. Informed consent for participation was obtained from study participants or their guardians.

3. Results

The study included 240 out of 250 patients consecutively admitted to the geriatric ward, with 177 (73.8%) being women and 193 (80.4%) aged 75 and above. The average age for men was 81 (±7.65) years, and for women, it was 80 (±6.27) years, with no statistically significant difference between the genders. The median 25(OH)D concentration for the entire study group was 22.95 (IQR 13.7–33.0) ng/mL. Breaking down by gender, the median VD concentration for women was 23 (IQR 14.0–34.0) ng/mL, and for men, it was 22 (IQR 13.0–32.0) ng/mL, with no significant differences observed in VD concentration between the sexes (Table 1). In terms of age groups, patients aged 60–74 showed a median VD concentration of 24 (IQR 14.0–33.0) ng/mL, while those aged 75 and above had a median VD concentration of 19.8 (IQR 13.0–30.0) ng/mL ($p > 0.05$).

Table 1. Vitamin D concentration strata in the sex groups.

25(OH)D Concentration [ng/mL]	Women (n = 177)		Men (n = 63)		Total (n = 240)		p-Value
	n	%	n	%	n	%	
<0; 10	30	16.9	6	9.5	36	15.0	
<10; 20	50	28.2	22	34.9	72	30.0	
<20; 30	40	22.6	14	22.2	54	22.5	0.73
<30; 50	43	24.3	19	30.2	62	25.8	
<50; 100	14	7.9	2	3.2	16	6.7	

A significant difference in VD concentration was observed among BMI categories (p = 0.048) (Figure 2a). The highest median VD concentration was found in the overweight group (BMI 25–29.99 kg/m^2) at 29 (IQR 16.0–38.0) [ng/mL]. Patients classified as obese (BMI 30 kg/m^2 or more) had a median VD concentration of 22 (IQR 15.00–29.60), while those with a normal BMI level (<25 kg/m^2) had a median of 21 (IQR 11.00–32.50) [ng/mL].

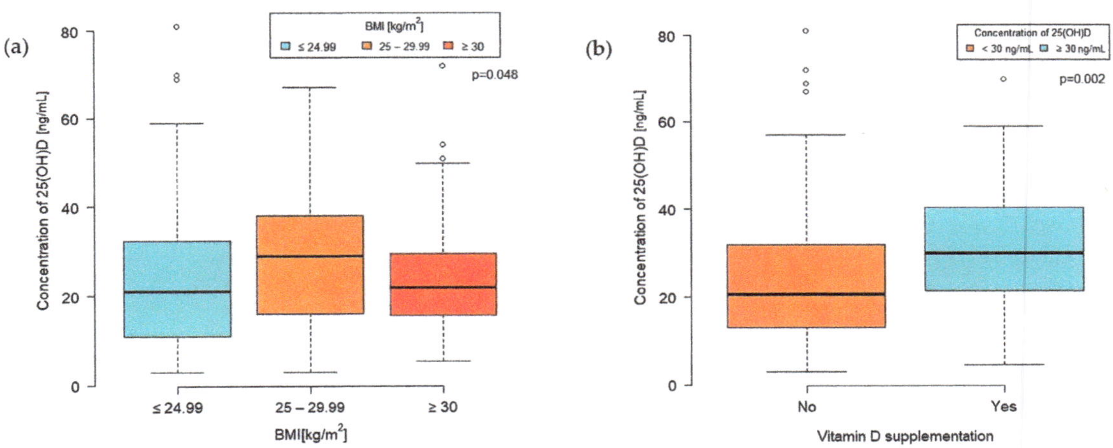

Figure 2. Concentration of VD by BMI group (**a**) and VD supplementation (**b**).

The prevalence of VD inadequacy (25(OH)D < 30 ng/mL) in the study group was 67.5% (162 patients). The deficiency in 15% of cases was profound, with a VD level of <10 ng/mL. The percentage of VD inadequacy was similar between the sexes (Figure 3a). Individuals aged 60–74 had a slightly higher prevalence of VD inadequacy than those aged 75 and older (Figure 3d). However, there were no statistically significant differences in VD inadequacy between sex and age groups.

The prevalence of VD inadequacy in the study group was also not associated with the place of residence (p = 0.753) (Figure 3b). However, those living in the countryside exhibited the highest average VD concentrations, with a median VD concentration of (IQR: 17.00–33.60). Similar median values of VD concentrations were observed among those living in large cities (Me: 22; IQR: 13.00–33.00) and towns (Me: 22.50; IQR: 15.00–37.00).

A parameter significantly associated with the prevalence of VD inadequacy was BMI (p = 0.01). Patients classified as overweight (BMI 25–29.99 kg/m^2) had the lowest prevalence of VD inadequacy, with a rate of 54.69% (Figure 3c). The prevalence of inadequacy among patients with a normal BMI (18.5–24.99 kg/m^2) was 66.67%. Patients diagnosed with obesity (BMI $\geq$ 30 kg/m^2) showed the highest prevalence of VD inadequacy, equal to 76.24%. Calculating OR of VD inadequacy between patients diagnosed with obesity (BMI $\geq$ 30 kg/m^2) and those without obesity (BMI < 30 kg/m^2) resulted in

OR = 2.21 (95% CI: 1.24–3.95) (p = 0.0074). This indicates that the odds of VD inadequacy are 2.21 times higher in patients diagnosed with obesity than in the group without obesity.

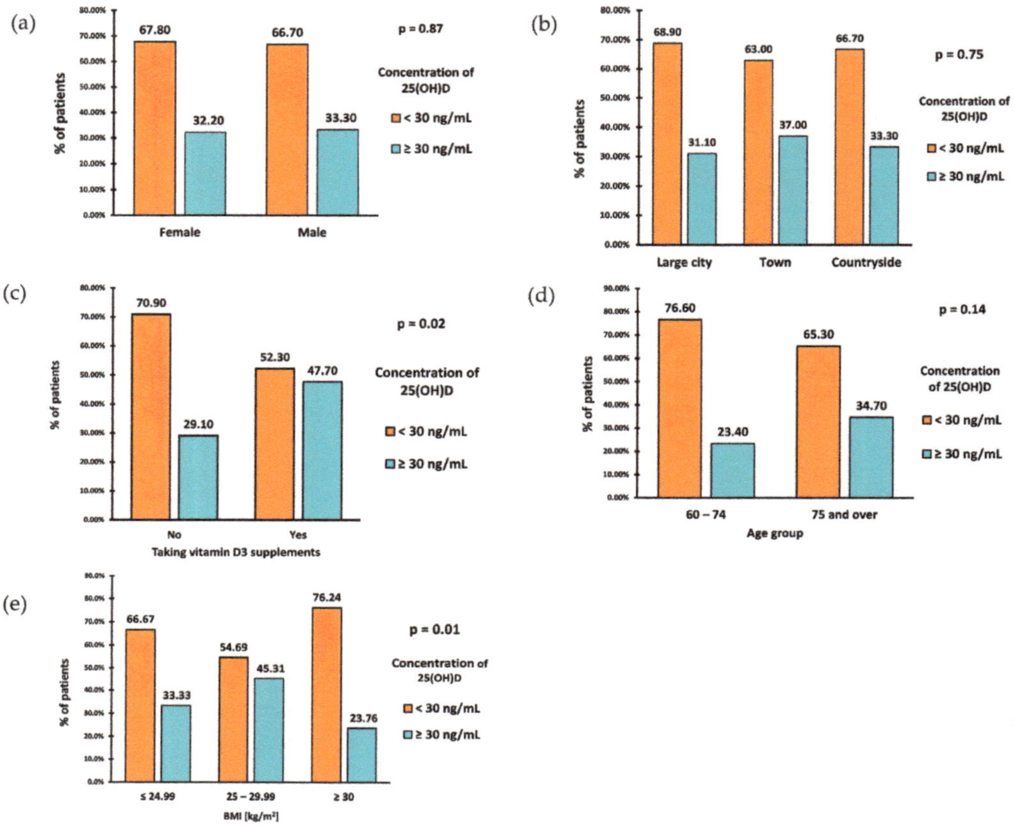

Figure 3. Prevalence of VD inadequacy by sex (**a**), place of residence (**b**), intake of vitamin D supplements (**c**), age (**d**), and BMI (**e**).

Furthermore, the concentration of 25-hydroxyvitamin D was significantly higher in the group taking VD supplements than those who did not (p = 0.002). The median VD concentration in the supplementation group was 29.80 (IQR 21.50–40.50) ng/mL, whereas in the nonsupplementation group, it was 20.50 (IQR 13.00–32.00) ng/mL (Figure 2b).

Only 44 patients (18.33%) reported taking VD supplements. Among the 196 (81.67%) who did not take supplements, 139 (70.9%) were found to have VD inadequacy, while 57 (29.10%) had proper levels (Figure 3c). Among those who claimed to take supplements, 23 (52.30%) had VD inadequacy, while 21 (47.70%) had proper levels. The statistical analysis demonstrated a significant association between taking VD supplements and experiencing inadequacy (p = 0.017). The odds of VD inadequacy in the nonsupplemented VD group were 2.43, while the odds of inadequacy in the supplemented group were 1.09. After calculating the odds ratio, the result was 2.23 (95% CI: 1.14–4.34) (p = 0.0187). This indicates that the odds of VD inadequacy in the nonsupplementing VD group are 2.23 times higher than in the supplementing VD group. We analyzed the relationship between the occurrence of insufficiency of vitamin D concentration (i.e., <30 ng/mL) and sex, age, place of residence, and BMI separately for the groups taking and not taking vitamin D supplements. The results were similar and presented together. There was no significant association between age, gender, and the use of VD supplements, respectively (Table 2). There was also no significant association between the place of residence and the use of VD supplements.

Table 2. Use of vitamin D3 supplements by sex and age.

	Taking Vitamin D3 Supplements		p-Value
	Yes *n* (%)	**No** *n* (%)	
Sex			
Woman (n = 177)	33(18.6)	144(81.4)	0.13
Man (n = 63)	11 (17.5)	52 (82.5)	
Age			
60–74 years (n = 47)	5 (10.6)	42 (89.4)	0.84
75+ years (n = 193)	39(20.2)	154 (79.8)	

In the multivariable logistic regression assessing the concurrent association between VD inadequacy, supplementation, and obesity diagnosis, the AUC was 0.625 (±0.03) (Table 3). It was found that by holding the obesity diagnosis constant, the odds of VD inadequacy increased by 128% (95% CI: 1.13–4.58) for patients not supplementing VD compared to supplementing VD. It was also found that by holding VD supplementing constant, the odds of VD inadequacy increased by 112% (95% CI: 1.19–3.86) for obese patients compared to nonobese patients.

Table 3. Multivariate logistic regression analysis—association of vitamin D inadequacy with VD supplementation and obesity.

Coefficients	Estimate	Std. Error	Odds Ratios	95% CI	p-Value
(Intercept)	−1.309	0.2462	0.27	0.16–0.43	<0.001
VD supplementation (no)	0.8228	0.3542	2.28	1.13–4.58	0.02
Obesity (BMI $\geq$ 30 kg/m^2)	0.7527	0.2999	2.12	1.19–3.86	0.01

4. Discussion

Our research confirmed that in older patients admitted to the Department of Geriatrics, VD concentration is below the appropriate value in most cases. The concentration of 25(OH)D ranged from 3 to 81 ng/mL, but it was within the optimal level (i.e., within the range $\geq$ 30–50 ng/mL) in only 25% of patients (Table 1). Only 6.7% had a high level, i.e., $\geq$50 ng/mL, whereas potentially toxic levels, i.e., >100 ng/mL, were not found. Therefore, as our study shows, most older patients of the geriatrics ward, regardless of age, sex, or place of residence, are inadequate in VD. Older people are at risk of VD insufficiency due to age-related impaired skin biosynthesis, more frequently observed inadequate absorption and metabolism [1]. According to some studies, nursing home residents and older inpatients are at the highest risk for VD deficiency [29].

In the nationwide PolSenior2 study in 2018–2019 among community-dwelling older people in Poland, severe VD deficiency was found in 11.5% of respondents, deficiency in 44.8%, and a suboptimal level in 30.9%. Only 12.7% of the subjects had a concentration of VD within the optimal range, defined as $\geq$30 ng/mL [22]. Compared to these results, the VD status of geriatrics ward patients participating in our study in 2019 was better, probably due to their frequent use of medical advice, including the geriatrics outpatient clinic, and the opportunities to receive recommendations on proper vitamin D3 supplementation.

A study conducted in 2019 in Europe and the Middle East indicates that low VD levels may affect up to 40% of the general population with 12.5% experiencing a severe deficiency [21]. Seniors and nursing home residents had worse results than the general population. It is also worth emphasizing that the percentage of people with a deficiency is relatively low in Scandinavian countries despite unfavorable geographical conditions (due to the geographical latitude, skin synthesis is significantly limited) [17]. Between 6.6% and 33.6% of the population of northern Europe has VD < 20 ng/mL, while in Western Europe, this prevalence range varies from 27.2% to even 61.4%. The most crucial factor here is widely used preventive measures such as VD supplementation, consumption of products rich in it (e.g., cod liver oil), and fortifying food with this vitamin. The estimated supply of VD in northern countries is about 10 mcg/day on average, while in Western European countries, it ranges from 1.5 mcg/day to about 5 mcg/day [21]. Unfortunately, there is no standardized data on this subject for Eastern Europe. Still, the deficiency affects a more significant percentage of the population than in Northern or Western European countries, and the average concentration is usually below 20 ng/mL.

In cross-sectional studies, the lowest average value was observed in Germany (39.50 $\pm$ 7.50 nmol/L) (15.82 $\pm$ 3.00 ng/mL), while the highest average concentrations were observed in Austria (56.50 $\pm$ 30.70 nmol/L) (22.63 $\pm$ 12.29 ng/mL) and Spain (50.70 $\pm$ 29.2 nmol/L) (20.31 $\pm$ 11.69 ng/mL) [25,26]. The incidence of VD deficiency is estimated from 30% of the elderly population in Germany to as high as 60% among Eastern European countries [21,30,31]. Retrospective research from the United Kingdom, analyzing data from 2005 to 2015, states that approximately 30% of surveyed women and men over the age of 65 years have a confirmed deficiency [32].

Therefore, despite the commonly available knowledge on the importance of VD for human health and published international recommendations on preventing its deficits, VD deficiency is a global public health concern. In 2009, the Polish standardized guidelines for VD supplementation were established [33]. The recommended supplementation dose for adults ranged between 800 and 1000 IU/day, with no specific recommendations for individuals aged over 75 years. In 2013, new guidelines for the Central European population were published, introducing many changes to the reference values. The proposed optimal concentration was set between 30 and 50 ng/mL with a more stringent deficiency threshold of 20 ng/mL [34]. The adult supplementation dose range has been increased, setting it at 800–2000 IU/day. However, there were still no specific recommendations for seniors over 75 years. The subsequent change occurred in 2018, with the next update of recommendations [16]. It suggested that older seniors should use a higher dose, ranging between 2000 and 4000 IU/day, depending on their weight and initial VD concentration in plasma. It also distinguished a deficiency in the range of 10–20 ng/mL, and values below 10 ng/mL were classified as a severe deficiency. The latest guidelines from 2023 still maintain the previously established reference values, only raising the lower dose limit for supplementation to 1000 IU/day instead of 800 IU/day [15]. Nevertheless, despite the prevention standards having been in force for many years, instances of severe VD deficiency persist in this population. Our research has shown that it is essential to consider a higher supplementation dose for obese individuals. The presence of obesity more than doubles the likelihood of an inadequate VD concentration, even with controlled supplementation. The reason for the increased need for VD augmentation in people with obesity is not immediately apparent. Increased VD clearance or impaired VD synthesis does not seem to be the correct explanation, as the values do not differ between the obese population and that of normal-weight individuals. Therefore, at this point, the most relevant factor seems to be excessive "dilution" due to increased mass and tissue volume in obese individuals [35].

The use of vitamin, mineral, and fish oil supplements is common among older people living in the community outside of care institutions, and studies suggest an increase in their consumption, particularly of calcium and VD, over the years [36]. According to current guidelines, VD supplementation in the senior population should be maintained year-round [15,36]. Despite the fact that recommendations for preventive VD supplementation have been in force in Poland for about 15 years, the prevalence of inadequacy and deficiency remains high, suggesting that the number of seniors adhering to these recommendations is meager. In the nationwide PolSenior2 study conducted in 2018–2019, only 16.3% of women and 6.4% of men over 60 reported taking vitamin D3 supplements [22]. Consequently, in both groups—the general population of seniors in the PolSenior2 study and those hospitalized in our study—most elderly subjects had VD inadequacy.

Based on the available literature, improving the VD status among older adults will require a multifaceted approach. One crucial step is to intensify the focus on educating the community, particularly seniors. The potential challenge lies in the low compliance observed among older patients, especially when they are physically and cognitively impaired. It then becomes imperative to educate their caregivers properly. This issue has been identified in care institutions, and adequate education and training of staff responsible for nursing home residents has demonstrated improvement in addressing VD deficiencies [37].

Research conducted in Poland highlights the importance of extending educational efforts to include physicians, particularly family physicians, as they are the first line of contact for patients. The study titled "Vitamin D Knowledge, Attitudes, and Practices of Polish Medical Doctors" analyzed doctors' knowledge of VD in the context of their clinical practice and personal application of the recommendations. The study involved 701 medical doctors from various specialties, predominantly women, working in diverse settings. The findings revealed that approximately 25% of the participants had their VD levels assessed, and 62% were found to have low levels. Although 82% reported that they provide information to patients about recommendations regarding supplementation, only around 40% did so for most patients. Alarmingly, 18% of surveyed doctors never discussed recommendations concerning VD supplementation with their patients [38]. This is crucial because both the study conducted in the geriatric ward and the nationwide PolSenior2 study demonstrate that VD supplementation significantly improved VD status. In both studies, a majority of seniors receiving VD preparations achieved optimal levels of 25(OH)D.

According to some authors, maintaining an optimal VD concentration requires a combination of education, supplementation, and regular monitoring of VD levels [39]. Boluses of higher doses of VD3 may improve compliance with prophylactic supplementation and deficiency therapy recommendations. The extended serum half-life (4–6 weeks) and tissue storage of the circulating metabolite of VD allow for less frequent administration, such as every other day, twice a week, or once a week. Intervention studies indicate that preparations containing higher VD doses can help achieve and maintain the appropriate, stable level of 25(OH)D. However, if the level remains insufficient, patients should avoid excessively escalating VD doses in boluses. Instead, in such cases, the intervals between doses should be shortened [40].

Another potential measure can be a more significant enrichment of commonly consumed food products like margarine, milk, or eggs with VD [41,42]. As the modeling results for Canada's population suggest, while this policy could improve VD levels for individuals not taking VD supplements, the overwhelming majority of Canadians (76%) would still fall below the estimated average requirement (10 mg/day or 400 IU/day). Therefore, it remains crucial to promote VD supplementation to address population-level VD inadequacies [43,44].

The high quality of VD supplements is also very important [45]. In the context of Poland, it is important to note that only a few of the VD preparations available in the market have the status of a "drug," ensuring a reliably assessed content of cholecalciferol in each tablet. Some older people may take various multivitamins with differing VD content and possibly varying bioavailability, which could be a contributing factor [46]. Efforts should be made to ensure that the VD supplements consumed by older people are of high quality.

When interpreting the results of the presented study, it is important to consider its limitations, which constrain the extent of the conclusions that can be drawn. This is a single-centered study concentrating on VD intake upon admission to the department, with no inquiry into past or periodic supplementation. Information about the type of supplement (whether VD or multivitamin preparations) and the dosage taken by patients is not available. This study neither assesses the patient's/caregiver's knowledge on this topic nor the reasons for supplement intake. Additionally, the analysis did not include questions about sun exposure and consumption of VD-enriched foods. It is important to perceive this analysis as a preliminary study, indicating the need for further investigation.

5. Conclusions

The conducted study showed that despite recommendations for VD supplementation having been in place for over a decade, the percentage of seniors admitted to the Department of Geriatrics with VD deficiency and inadequacy remains alarmingly high. This emphasizes the urgent need for further actions to address this concerning trend. These actions should include educational initiatives targeting older adults and their caregivers, along with enhancing the knowledge of physicians caring for seniors. An assessment of activities promoting VD supplementation should be conducted to optimize their effectiveness. Additionally, research on available VD supplements in the market is crucial, considering that only select preparations in Poland hold "drug" status and meet quality standards. Modifications of therapy methods, using preparations containing higher doses of VD, may help to improve compliance by older adults. Improvement at the population level may also be brought about by promoting the availability of food products fortified with VD or increasing the degree of this food enrichment.

Author Contributions: Conceptualization, M.W., M.A.L., A.W. and Z.B.W.; methodology, M.Ś. and Z.B.W.; investigation, M.Ś. and Z.B.W.; software, M.W. and Z.B.W.; validation, M.W., M.A.L., A.W., J.R., and Z.B.W.; formal analysis, M.W. and Z.B.W.; data curation, M.W. and Z.B.W.; writing—original draft preparation, M.W., M.A.L., A.W., and Z.B.W.; writing—review and editing, M.W., M.A.L., A.W., and Z.B.W.; visualization, M.W.; supervision, Z.B.W.; funding acquisition, M.Ś., Z.B.W. All authors have read and agreed to the published version of the manuscript.

Funding: The research was funded by the Medical University of Bialystok under grant number N/ST/ZB/15/001/3301, awarded to M.S. The financial sponsor had no involvement in the design, execution, analysis, or interpretation of the data. We presented this paper, including interim findings, at the 19th International Congress of the European Geriatric Medicine Society (20–22 September 2023, Helsinki, Finland), as a poster presentation (*Eur. Geriatr. Med.* **2023**, *14* (Suppl. 1), S319) [47].

Institutional Review Board Statement: The study was conducted in adherence to the Declaration of Helsinki and received approval from the Ethics Committee of the Medical University of Bialystok (no R-I-002/326/2017, 28 September 2017).

Informed Consent Statement: Informed consent was obtained from all subjects involved in the study.

Data Availability Statement: The data presented in this study are available on request from the corresponding author. The data are not publicly available due to privacy and ethical considerations.

Acknowledgments: Figure 1 was created using GIMP 2.10 software. Four anonymous peer reviewers are kindly acknowledged for their comments and suggestions, which helped improve the manuscript.

Conflicts of Interest: The authors declare no conflicts of interest.

Abbreviations

VD—Vitamin D; MARRS—membrane-associated rapid response steroid; VDR—Vitamin D receptors; UV-B—ultraviolet B radiation; DBP—vitamin D binding protein; CGA—comprehensive geriatric assessment; BMI—Body Mass Index; M—mean; SD—standard deviation; Me—median; IQR—interquartile range; OR—odds ratio; CI—confidence interval.

References

1. Kupisz-Urbańska, M.; Łukaszkiewicz, J.; Marcinowska-Suchowierska, E. Vitamin D in Elderly. In *Vitamin D*; IntechOpen: Rijek Croatia, 2021. [CrossRef]
2. Yaman, A.E.; Ceylan, U.S. Effects of Vitamin D Levels on Long-Term Coronary Events in Patients with Proven Coronary Arter Disease: Six-Year Follow-Up. *J. Clin. Med.* **2023**, *12*, 6835. [CrossRef]
3. Ao, T.; Kikuta, J.; Ishii, M. The Effects of Vitamin D on Immune System and Inflammatory Diseases. *Biomolecules* **2021**, *11*, 162 [CrossRef]

4. Pludowski, P.; Grant, W.B.; Konstantynowicz, J.; Holick, M.F. Editorial: Classic and Pleiotropic Actions of Vitamin D. *Front. Endocrinol.* **2019**, *10*, 341. [CrossRef] [PubMed]
5. Lason, W.; Jantas, D.; Leskiewicz, M.; Regulska, M.; Basta-Kaim, A. The Vitamin D Receptor as a Potential Target for the Treatment of Age-Related Neurodegenerative Diseases Such as Alzheimer's and Parkinson's Diseases: A Narrative Review. *Cells* **2023**, *12*, 660. [CrossRef] [PubMed]
6. Martin-Moreno, J.M.; Martin-Gorgojo, A. "The Role of Vitamin D in Cancer Prevention": Some New Clues on a Fascinating Subject. *Nutrients* **2023**, *15*, 2560. [CrossRef]
7. Haykal, T.; Samji, V.; Zayed, Y.; Gakhal, I.; Dhillon, H.; Kheiri, B.; Kerbage, J.; Veerapaneni, V.; Obeid, M.; Danish, R.; et al. The role of vitamin D supplementation for primary prevention of cancer: Meta-analysis of randomized controlled trials. *J. Community Hosp. Intern. Med. Perspect.* **2019**, *9*, 480–488. [CrossRef] [PubMed]
8. Arayici, M.E.; Basbinar, Y.; Ellidokuz, H. Vitamin D Intake, Serum 25-Hydroxyvitamin-D (25(OH)D) Levels, and Cancer Risk: A Comprehensive Meta-Meta-Analysis Including Meta-Analyses of Randomized Controlled Trials and Observational Epidemiological Studies. *Nutrients* **2023**, *15*, 2722. [CrossRef]
9. Holick, M.F. Resurrection of vitamin D deficiency and rickets. *J. Clin. Investig.* **2006**, *116*, 2062–2072. [CrossRef]
10. Tabaei, S.; Motallebnezhad, M.; Tabaee, S.S. Vitamin D Receptor (VDR) Gene Polymorphisms and Risk of Coronary Artery Disease (CAD): Systematic Review and Meta-analysis. *Biochem. Genet.* **2021**, *59*, 813–836. [CrossRef]
11. Botelho, J.; Machado, V.; Proenca, L.; Delgado, A.S.; Mendes, J.J. Vitamin D Deficiency and Oral Health: A Comprehensive Review. *Nutrients* **2020**, *12*, 1471. [CrossRef]
12. Ofoedu, C.E.; Iwouno, J.O.; Ofoedu, E.O.; Ogueke, C.C.; Igwe, V.S.; Agunwah, I.M.; Ofoedum, A.F.; Chacha, J.S.; Muobike, O.P.; Agunbiade, A.O.; et al. Revisiting food-sourced vitamins for consumer diet and health needs: A perspective review, from vitamin classification, metabolic functions, absorption, utilization, to balancing nutritional requirements. *PeerJ* **2021**, *9*, e11940. [CrossRef] [PubMed]
13. Liu, L.; Jin, R.; Hao, J.; Zeng, J.; Yin, D.; Yi, Y.; Zhu, M.; Mandal, A.; Hua, Y.; Ng, C.K.; et al. Consumption of the Fish Oil High-Fat Diet Uncouples Obesity and Mammary Tumor Growth through Induction of Reactive Oxygen Species in Protumor Macrophages. *Cancer Res.* **2020**, *80*, 2564–2574. [CrossRef] [PubMed]
14. Serrano, D.; Bellerba, F.; Johansson, H.; Macis, D.; Aristarco, V.; Accornero, C.A.; Guerrieri-Gonzaga, A.; Trovato, C.M.; Zampino, M.G.; Sale, E.O.; et al. Vitamin D Supplementation and Adherence to World Cancer Research Fund (WCRF) Diet Recommendations for Colorectal Cancer Prevention: A Nested Prospective Cohort Study of a Phase II Randomized Trial. *Biomedicines* **2023**, *11*, 1766. [CrossRef]
15. Pludowski, P.; Kos-Kudla, B.; Walczak, M.; Fal, A.; Zozulinska-Ziolkiewicz, D.; Sieroszewski, P.; Peregud-Pogorzelski, J.; Lauterbach, R.; Targowski, T.; Lewinski, A.; et al. Guidelines for Preventing and Treating Vitamin D Deficiency: A 2023 Update in Poland. *Nutrients* **2023**, *15*, 695. [CrossRef] [PubMed]
16. Rusińska, A.; Płudowski, P.; Walczak, M.; Borszewska-Kornacka, M.K.; Bossowski, A.; Chlebna-Sokół, D.; Czech-Kowalska, J.; Dobrzańska, A.; Franek, E.; Helwich, E.; et al. Vitamin D Supplementation Guidelines for General Population and Groups at Risk of Vitamin D Deficiency in Poland-Recommendations of the Polish Society of Pediatric Endocrinology and Diabetes and the Expert Panel With Participation of National Specialist Consultants and Representatives of Scientific Societies-2018 Update. *Front. Endocrinol.* **2018**, *9*, 246. [CrossRef]
17. Czerwińska, A.; Czuchraj, W. Estimations of the Erythemal UV Doses and the Amount of the Sun-Synthesized Vitamin D by Adults during the Cruise to Spitsbergen–Polar Measurement Campaign (2–21 July 2017). *Atmosphere* **2021**, *12*, 474. [CrossRef]
18. Dominguez, L.J.; Farruggia, M.; Veronese, N.; Barbagallo, M. Vitamin D Sources, Metabolism, and Deficiency: Available Compounds and Guidelines for Its Treatment. *Metabolites* **2021**, *11*, 255. [CrossRef]
19. Amrein, K.; Scherkl, M.; Hoffmann, M.; Neuwersch-Sommeregger, S.; Kostenberger, M.; Tmava Berisha, A.; Martucci, G.; Pilz, S.; Malle, O. Vitamin D deficiency 2.0: An update on the current status worldwide. *Eur. J. Clin. Nutr.* **2020**, *74*, 1498–1513. [CrossRef]
20. Cashman, K.D.; Dowling, K.G.; Skrabakova, Z.; Gonzalez-Gross, M.; Valtuena, J.; De Henauw, S.; Moreno, L.; Damsgaard, C.T.; Michaelsen, K.F.; Molgaard, C.; et al. Vitamin D deficiency in Europe: Pandemic? *Am. J. Clin. Nutr.* **2016**, *103*, 1033–1044. [CrossRef]
21. Lips, P.; Cashman, K.D.; Lamberg-Allardt, C.; Bischoff-Ferrari, H.A.; Obermayer-Pietsch, B.; Bianchi, M.L.; Stepan, J.; El-Hajj Fuleihan, G.; Bouillon, R. Current vitamin D status in European and Middle East countries and strategies to prevent vitamin D deficiency: A position statement of the European Calcified Tissue Society. *Eur. J. Endocrinol.* **2019**, *180*, P23–P54. [CrossRef]
22. Błędowski, P.; Grodzicki, T.; Mossakowska, M.; Zdrojewski, T.; Społecznego, I.G.; i Dydaktyki, Z.P. *PolSenior2—Badanie Poszczególnych Obszarów Stanu Zdrowia Osób Starszych, w Tym Jakości życia Związanej ze Zdrowiem*; Gdański Uniwersytet Medyczny: Gdańsk, Poland, 2021.
23. Wierucki, Ł.; Kujawska-Danecka, H.; Mossakowska, M.; Grodzicki, T.; Błędowski, P.; Chudek, J.; Kostka, T.; Więcek, A.; Hajduk, A.; Bandosz, P.; et al. Health status and its socio-economic covariates in the older population in Poland—The assumptions and methods of the nationwide, cross-sectional PolSenior2 survey. *Arch. Med. Sci.* **2022**, *18*, 92–102. [CrossRef] [PubMed]

24. Wyskida, M.; Owczarek, A.; Szybalska, A.; Brzozowska, A.; Szczerbowska, I.; Wieczorowska-Tobis, K.; Puzianowska-Kuźnicka, M.; Franek, E.; Mossakowska, M.; Grodzicki, T.; et al. Socio-economic determinants of vitamin D deficiency in the older Polish population: Results from the PolSenior study. *Public Health Nutr.* **2018**, *21*, 1995–2003. [CrossRef] [PubMed]
25. Charkiewicz, M.; Wojszel, Z.B.; Kasiukiewicz, A.; Magnuszewski, L.; Wojszel, A. Association of Chronic Heart Failure with Frailty, Malnutrition, and Sarcopenia Parameters in Older Patients-A Cross-Sectional Study in a Geriatric Ward. *J. Clin. Med.* **2023**, *12*, 2305. [CrossRef] [PubMed]
26. Wojszel, Z.B. What Serum Sodium Concentration Is Suggestive for Underhydration in Geriatric Patients? *Nutrients* **2020**, *12*, 496. [CrossRef] [PubMed]
27. Wojszel, Z.B. Impending Low Intake Dehydration at Admission to A Geriatric Ward- Prevalence and Correlates in a Cross-Sectional Study. *Nutrients* **2020**, *12*, 398. [CrossRef] [PubMed]
28. WHO Consultation. *Obesity: Preventing and Managing the Global Epidemic. Report of a WHO Consultation*; World Health Organization: Geneva, Switzerland, 2000; Volume 894, pp. 1–253. Available online: https://books.google.pl/books?id=AvnqOsqv9doC&printsec=frontcover&hl=pl&source=gbs_ge_summary_r&cad=0#v=onepage&q&f=false (accessed on 5 December 2023).
29. Griffin, T.P.; Wall, D.; Blake, L.; Griffin, D.G.; Robinson, S.M.; Bell, M.; Mulkerrin, E.C.; O'Shea, P.M. Vitamin D Status of Adults in the Community, in Outpatient Clinics, in Hospital, and in Nursing Homes in the West of Ireland. *J. Gerontol. A Biol. Sci. Med. Sci.* **2020**, *75*, 2418–2425. [CrossRef] [PubMed]
30. Richter, K.; Breitner, S.; Webb, A.R.; Huth, C.; Thorand, B.; Kift, R.; Linseisen, J.; Schuh, A.; Kratzsch, J.; Mielck, A.; et al. Influence of external, intrinsic and individual behaviour variables on serum 25(OH)D in a German survey. *J. Photochem. Photobiol. B* **2014**, *140*, 120–129. [CrossRef]
31. van Schoor, N.; Lips, P. Global Overview of Vitamin D Status. *Endocrinol. Metab. Clin. N. Am.* **2017**, *46*, 845–870. [CrossRef]
32. Crowe, F.L.; Jolly, K.; MacArthur, C.; Manaseki-Holland, S.; Gittoes, N.; Hewison, M.; Scragg, R.; Nirantharakumar, K. Trends in the incidence of testing for vitamin D deficiency in primary care in the UK: A retrospective analysis of The Health Improvement Network (THIN), 2005–2015. *BMJ Open* **2019**, *9*, e028355. [CrossRef]
33. Charzewska, J.; Chlebna-Sokół, D.; Chybicka, A.; Czech-Kowalska, J.; Dobrzańska, A.; Helwich, E.; Imiela, J.R.; Karczmarewicz, E.; Ksiazyk, J.B.; Lewiński, A.; et al. Polskie zalecenia dotyczące profilaktyki niedoborów witaminy D—2009 [Prophylaxis of vitamin D deficiency—Polish recommendation 2009]. *Ginekol Pol.* **2010**, *81*, 149–153.
34. Płudowski, P.; Karczmarewicz, E.; Bayer, M.; Carter, G.; Chlebna-Sokół, D.; Czech-Kowalska, J.; Dębski, R.; Decsi, T.; Dobrzańska, A.; Franek, E.; et al. Practical guidelines for the supplementation of vitamin D and the treatment of deficits in Central Europe—Recommended vitamin D intakes in the general population and groups at risk of vitamin D deficiency. *Endokrynol. Pol.* **2013**, *64*, 319–327. [CrossRef] [PubMed]
35. Walsh, J.S.; Bowles, S.; Evans, A.L. Vitamin D in obesity. *Curr. Opin. Endocrinol. Diabetes Obes.* **2017**, *24*, 389–394. [CrossRef] [PubMed]
36. Savikko, N.; Pitkälä, K.H.; Laurila, J.V.; Suominen, M.H.; Tilvis, R.S.; Kautiainen, H.; Strandberg, T.E. Secular trends in the use of vitamins, minerals and fish-oil products in two cohorts of community-dwelling older people in Helsinki--population-based surveys in 1999 and 2009. *J. Nutr. Health Aging* **2014**, *18*, 150–154. [CrossRef] [PubMed]
37. Feehan, O.; Magee, P.J.; Pourshahidi, L.K.; Armstrong, D.J.; McSorley, E.M. Vitamin D deficiency in nursing home residents: A systematic review. *Nutr. Rev.* **2023**, *81*, 804–822. [CrossRef] [PubMed]
38. Zgliczyński, W.S.; Rostkowska, O.M.; Sarecka-Hujar, B. Vitamin D Knowledge, Attitudes and Practices of Polish Medical Doctors. *Nutrients* **2021**, *13*, 2443. [CrossRef] [PubMed]
39. Lee, C.J.; Kim, S.S.; Suh, W.Y.; Kim, J.S.; Jung, J.G.; Yoon, S.J.; Seo, Y.R.; Yang, H.J. The Effect of Education and Vitamin D Supplementation on the Achievement of Optimal Vitamin D Level in Korean Postmenopausal Women. *J. Bone Metab.* **2019**, *26*, 193–199. [CrossRef] [PubMed]
40. Välimäki, V.V.; Löyttyniemi, E.; Pekkarinen, T.; Välimäki, M.J. How well are the optimal serum 25OHD concentrations reached in high-dose intermittent vitamin D therapy? a placebo-controlled study on comparison between 100 000 IU and 200 000 IU of oral D3 every 3 months in elderly women. *Clin. Endocrinol.* **2016**, *84*, 837–844. [CrossRef]
41. Sengupta, S.; Christensen, T.; Ravn-Haren, G.; Andersen, R. Vitamin D Food Fortification Strategies on Population-Based Dietary Intake Data Using Mixed-Integer Programming. *Foods* **2023**, *12*, 698. [CrossRef]
42. Vičič, V.; Mikuš, R.P.; Kugler, S.; Geršak, K.; Osredkar, J.; Kukec, A. Vitamin D Fortification of Eggs Alone and in Combination with Milk in Women Aged 44–65 Years: Fortification Model and Economic Evaluation. *Zdr. Varst.* **2023**, *62*, 30–38. [CrossRef]
43. Vatanparast, H.; Longworth, Z.L. How does Canada's new vitamin D fortification policy affect the high prevalence of inadequate intake of the vitamin? *Appl. Physiol. Nutr. Metab.* **2023**, *48*, 870–875. [CrossRef]
44. Dunlop, E.; James, A.P.; Cunningham, J.; Rangan, A.; Daly, A.; Kiely, M.; Nowson, C.A.; Adorno, P.; Atyeo, P.; Black, L.J. Vitamin D Fortification of Milk Would Increase Vitamin D Intakes in the Australian Population, but a More Comprehensive Strategy Is Required. *Foods* **2022**, *11*, 1369. [CrossRef] [PubMed]
45. Chugh, P.K.; Dabas, A. Price Dispersion of Vitamin D Supplements Over Time: An Initiative for Prescriber Education. *Indian J. Endocrinol. Metab.* **2021**, *25*, 142–147. [CrossRef] [PubMed]

46. Jittat, N.; Pongpirul, K.; Tepwituksakit, B.; Iammaleerat, P.; Heath, J.; Lungchukiet, P.; Taechakraichana, N.; Charukitpipat, A. Effects of Oral Multi-Vitamin Multi-Mineral Supplement Formulations on Laboratory Outcomes and Quality of Life: A Quasi-Experimental Study. *Front. Nutr.* **2022**, *9*, 889910. [CrossRef] [PubMed]
47. Christiaens, A.; Henrard, S.; Bonnet-Zamponi, D.; Zerah L. Abstracts of the 19th Congress of the European Geriatric Medicine Society: Live from Helsinki and Online, 20-22 September 2023. *Eur. Geriatr. Med.* **2023**, *14* (Suppl. S1), S319.

Disclaimer/Publisher's Note: The statements, opinions and data contained in all publications are solely those of the individual author(s) and contributor(s) and not of MDPI and/or the editor(s). MDPI and/or the editor(s) disclaim responsibility for any injury to people or property resulting from any ideas, methods, instructions or products referred to in the content.

nutrients MDPI

Brief Report

Vitamin D Status in Belgian Children: A Regional Study

Louise Van de Walle [1], Yvan Vandenplas [2], Jaan Toelen [1,3] and Anke Raaijmakers [4,5,*]

1 Department of Development and Regeneration, University of Leuven, 3000 Leuven, Belgium; jaan.toelen@uzleuven.be (J.T.)
2 KidZ Health Castle, UZ Brussel, Vrije Universiteit Brussel (VUB), 1090 Brussels, Belgium; yvan.vandenplas@uzbrussel.be
3 Department of General Pediatrics, University Hospitals Leuven, 3000 Leuven, Belgium
4 Department of Paediatric Nephrology, Sydney Children's Hospital Randwick, Sydney Children's Hospital Network, Randwick, NSW 2031, Australia
5 Randwick Clinical Campus, School of Women's and Children's Health, University of New South Wales, Randwick, NSW 2031, Australia
* Correspondence: anke.raaijmakers@health.nsw.gov.au

Abstract: *Background:* Vitamin D deficiency is the most frequent cause of impaired skeletal growth, and can lead to the development of nutritional rickets. The aim of this study was to evaluate the vitamin D status in a large group of children aged 0–18 years. *Methods:* We collected laboratory data on vitamin D levels from children who underwent blood sampling between 2014 and 2021. *Results:* We included 14,887 samples. In this group, 17.7% were vitamin D severely deficient (<12 ng/mL), 25.2% were insufficient (12–20 ng/mL), and another large proportion (28.3%) was borderline (20–30 ng/mL). Sufficient levels (>30 ng/mL) were met in 28.8% of children. We observed no association between gender and vitamin D status ($p = 0.132$). Adolescents aged 13–18 years ($n = 3342$) had the highest prevalence of severe vitamin D deficiency (24.9%). Vitamin D levels were higher in summer/autumn compared to winter/spring. *Conclusions:* Vitamin D deficiency/insufficiency has a high prevalence in children, mostly in children above 7 years of age. Many of these children (over 80%) do not meet the 30 ng/mL sufficiency threshold. It is essential that Belgian Health Authorities are aware of this high prevalence, as the current Belgian recommendation suggests ceasing vitamin D supplementation at the age of six. Additional research is required to investigate the consequences of our findings, and what specific approach is needed to achieve normal vitamin D levels in children aged 0 to 18 years.

Keywords: vitamin D; vitamin D status; vitamin D deficiency; rickets prevention

1. Introduction

Vitamin D plays a pivotal role in calcium and phosphate homeostasis, as well as in healthy bone metabolism, especially during growth [1]. In addition to the role of vitamin D in growth, it is believed that it might also plays a role in the regulation of the immune system [2,3].

Vitamin D is an umbrella term which comprises vitamin D3 (cholecalciferol) and vitamin D2 (ergocalciferol). Both forms are found in nature; vitamin D2 is derived from plants, whereas vitamin D3 is found in animal sources. However, the amount of vitamin D in natural food sources is low, and the contribution of dietary vitamin D is small or even negligible [4,5]. The main source of vitamin D3 comes from 7-dehydrocholesterol (previtamin D3) synthesis in the skin, in response to UVB rays. Solar UVB intensity is based on the season (higher radiation in summer) and geographical latitude (higher radiation in areas closer to the equator). The cutaneous synthesis of previtamin D3 cannot be detected above 50° geographical latitude from October to March [6], hence people in these areas (including Belgium) are at risk of low vitamin D levels, particularly in winter.

There is much debate about the optimal concentration of vitamin D in children. Research on pediatric vitamin D references is scarce [6]. Most studies on the optimal vitamin D levels are performed in adults [7]. Most experts agree that levels <12 ng/mL (30 nmol/L) represent severe deficiency, as these levels carry a high risk of rickets, even with an adequate calcium intake; today, most experts use 20–30 ng/mL (50 nmol/L) as the target value, which includes a safety margin [1,2,8,9]. For levels >30 ng/mL (75 nmol/L), the concentration of parathyroid hormones remains constant, which means that there is no secondary hyperparathyroidism. Some authors state that >30 ng/mL would be the logical 'optimal' vitamin D level [10], which is supported by the European Society of Pediatric Nephrology [9].

Vitamin D deficiency is the most frequent cause of impaired skeletal growth. Moreover, defective bone mineralization can lead to the development of rickets (or osteomalacia in older children) [11,12]. Globally, nutritional rickets, due to insufficient vitamin D and/or calcium, is the leading cause of pediatric bone disease [11,12]. However, the diagnosis of rickets is challenging, as the symptoms are insidious and difficult to recognize [8]. The use of surrogate markers is usually recommended [9]. Secondly, clinical research has demonstrated that vitamin D deficiency is associated with many acute and chronic illnesses, including calcium metabolism, several autoimmune diseases and cancers, type 2 diabetes mellitus, cardiovascular diseases, and potentially many more [4,7,13].

More than one billion humans worldwide are affected by vitamin D deficiency and insufficiency [1]. Additionally, a recent study performed in Belgium confirmed that vitamin D insufficiency and deficiency is still common in children, particularly at the end of winter [14]. In this study, we investigated the vitamin D status in children, visiting a regional hospital network (ZNA) in Antwerp, Belgium. We assessed vitamin D statuses and the factors associated with a poor status in a large group of children (0–18 years).

2. Materials and Methods

2.1. Data Collection and Subjects

Vitamin D data were retrospectively collected from the hospital central laboratory system, and personal data, barring sex and date of birth, were not including in the dataset. Children (aged 0–17 years and 364 days) who underwent a blood test including a vitamin D level in the ZNA hospitals network between 2014–2021 (n = 26,233) were retrieved. These years were chosen deliberately, as they represent the period after the consensus of vitamin D supplementation by the Flemish Society of Pediatrics [15]. Additional demographics (age, gender, and date of sampling) were also extracted. Individual data on vitamin D supplementation, diet, and level of sun exposure were not available, due to the retrospective and anonymous nature of the study.

We only included first-time samples in every patient, as subsequent samples are likely to be biased by comorbidities (unwell or chronically ill patients will have more samples) and/or treatment (low levels will trigger treatment). Samples taken at birth were considered maternal values and were excluded. Children having multiple ($\geq$2) vitamin D values measured in their sample or with missing information regarding their date of birth or gender were also excluded. The final sample size was n = 14,887 (Figure 1).

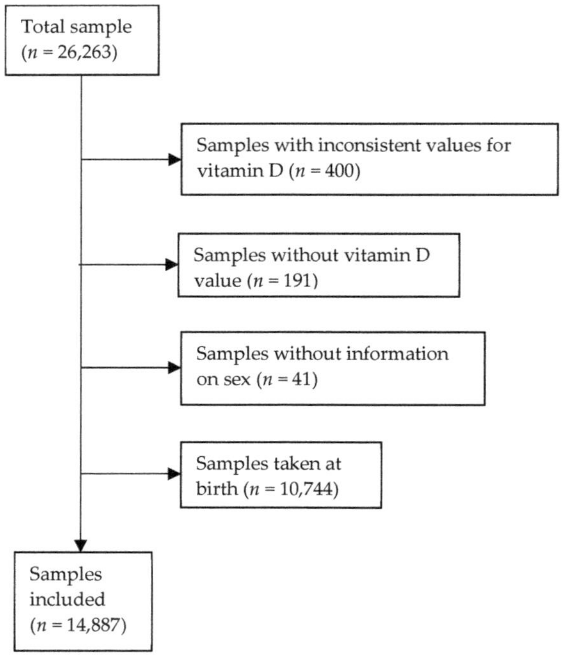

Figure 1. Overview of the total sample and the excluded and final included samples.

2.2. *Laboratory Assessment and Reference Values*

The quantitative determination of total 25-OH-vitamin D in the serum was performed by electrochemiluminescence binding assays (ECLIA) using Cobas E immunoassay analyzers (Roche Diagnostics, Mannheim). Vitamin D cut-offs were considered as per most used definitions of severely deficient, insufficient, borderline, and sufficient vitamin D levels of <12 ng/mL, 12–20 ng/mL, 20–30 ng/mL, and >30 ng/mL, respectively (Table 1) [1,2,8,9].

Table 1. Sample sizes by age and gender.

		Total (*n*)	Female (*n*)	Proportion (%)	Male (*n*)	Proportion (%)
Age in years	<1	2354 (15.8%)	1079	45.8%	1275	54.2%
	1–6	5070 (34.1%)	2348	46.3%	2722	53.7%
	7–12	4121 (27.7%)	2141	52.0%	1980	48.0%
	13–18	3342 (22.4%)	2016	60.3%	1326	39.7%
	Total	14,887 (100.0%)	7584	50.9%	7303	49.1%

2.3. *Statistical Analysis*

For database management and statistical analysis, we used SPSS Statistics for Macintosh (Version 28.0, Armonk, NY, USA). Data were measured at the ratio level. No significant outliers were identified based on the visual inspection of the created boxplots.

Frequency tables were calculated based on vitamin D levels in relation to age group and season. Differences in the frequency distributions of vitamin D status between categories were assessed using Pearson's Chi-Square test. Significant differences were considered if the p-value was <0.05. The following age categories were defined: <1 year, 1–6 years, 7–12 years, and 13–18 years. Seasons were defined according to the following months: spring (March, April, and May), summer (June, July, and August), autumn (September, October, and November) and winter (December, January, and February).

3. Results

3.1. Characteristics of the Cohort

Baseline characteristics are provided in Table 1. Males were slightly overrepresented in the youngest age group, while females were more prominent in the oldest group. The largest group were those aged between 1 and 6 years old (n = 5070, 34.1% of total sample size). Children up to 1 year old were the smallest group (n = 2354, 15.8% of total sample size).

3.2. Vitamin D Status

In our total sample, the distribution of the vitamin D serum concentration had a mean value of 23.94 ng/mL and a standard deviation of 12.11 ng/mL. The lowest level of concentration was 3.04 ng/mL. The highest level was 58.41 ng/mL.

Table 2 contains the distribution of vitamin D statuses in both absolute numbers and percentages within each age group for the total sample. Overall, 2635 (17.7%) were severely deficient, 3746 children (25.2%) were insufficient (12–20 ng/mL), and 4215 children (28.3%) were borderline (20–30 ng/mL). In total, 71.2% of the population did not meet the >30 ng/mL threshold, most prominent in the over-7 age groups (83.6% and 83.0% of children in the 7–12 and 13–18 age groups, respectively). There was no significant association between the vitamin D status and gender (Pearson Chi-Square: p = 0.132).

Table 2. Vitamin D status by age groups.

			Age				
			Years				Total
			<1	1–6	7–12	13–18	
Vitamin D status	Severely deficient <12 ng/mL	Count (n)	526	423	853	833	2635
		Percentage	22.3%	8.3%	20.7%	24.9%	17.7%
	Insufficient 12–20 ng/mL	Count (n)	547	875	1343	981	3746
		Percentage	23.2%	17.3%	32.6%	29.4%	25.2%
	Borderline 20–30 ng/mL	Count (n)	521	1483	1250	961	4215
		Percentage	22.1%	29.3%	30.3%	28.8%	28.3%
	Sufficient >30 ng/mL	Count (n)	760	2289	675	567	4291
		Percentage	32.3%	45.1%	16.4%	17.0%	28.8%
	Total	Count (n)	2354	5070	4121	3342	14,887

The prevalence of vitamin D insufficiency along with vitamin D deficiency was highest in 7–12-year-old children and adolescents (13–18 years), being 24.9% and 20.7%, respectively. In the youngest age group, 23.2% of the children had insufficient vitamin D levels, and 22.3% had severely deficient vitamin D levels. Overall, the groups of children above 7 years of age had the lowest vitamin D status.

In our study population, we were vigilant of seasonal variation. Vitamin D values measured during summer and autumn were significantly higher when compared to values measured in winter and spring. The proportion of vitamin D-sufficient children were highest in the summer and autumn months, as shown in Figure 2.

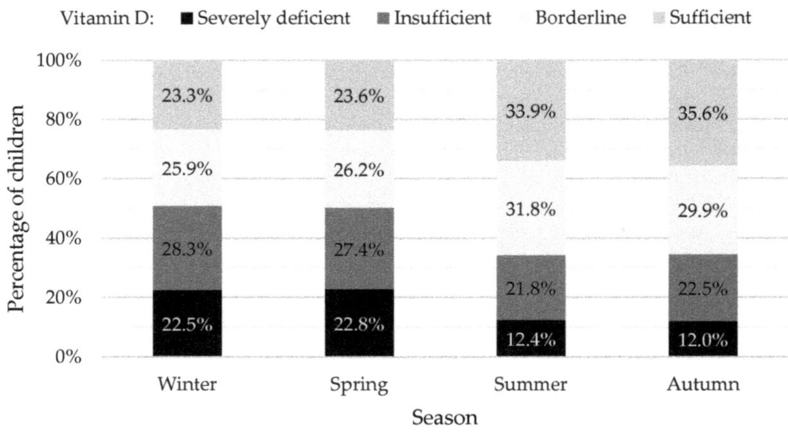

Figure 2. Percentages of vitamin D sufficient, insufficient, and deficient children.

Figure 2 is a graph showing the percentages of children who were vitamin D severely deficient, insufficient, borderline, and sufficient, according to the season. The percentages shown in black, dark grey, white, and light grey represent levels <12 ng/mL, 12–20 ng/mL, 20–30 ng/mL, and >30 ng/mL.

4. Discussion

Vitamin D deficiency is prevalent among Belgian children, with the highest prevalence in adolescents. The proportion of children with severely deficient vitamin D levels rises significantly with age, and reaches a peak before adolescence, where up to 55% of children have vitamin D levels below 20 ng/mL, and up to 84% have a vitamin D level that does not meet the sufficiency threshold (i.e., above 30 ng/mL). Additionally, our data confirm that at the end of winter vitamin D status is the lowest, when up to 77% of children have vitamin D levels below 30 ng/mL.

Our research shows results in line with studies in other European countries [6,16,17]. Furthermore, a recent study conducted in Belgium showed that up to half their population of adolescents have very low (<20 ng/mL) vitamin D levels [14]. We acknowledge that the cut-offs used in this study could be debated, and the stricter we would be, the more striking the results. The European Society for Pediatric Nephrology have developed recommendations for the evaluation, treatment, and prevention of vitamin D deficiency [9]. Their aim was to provide guidance on children with chronic kidney disease, but their systematic approach covered healthy children too. They recommend using levels above 30 ng/mL as sufficient [9], and our study concurred.

The reasons for the high prevalence of low vitamin D levels could be plenty. Firstly, (the lack of) supplements could play a role. The administration of 400 IU/day to babies during their first year of life, regardless of season, diet, or other potentially influencing factors, is a generally accepted recommendation in Europe to prevent nutritional rickets in babies. Despite this, our results show that only a third (32.3%) of the infants had sufficient vitamin D levels. These results are in line with previous studies, which showed that Belgian babies often have insufficient vitamin D levels [14,18].

Secondly, children with chronic diseases or conditions such as obesity do need extra vitamin D, especially in winter [1,3,7,9,19]. Vitamin D deficiency in obese children is mostly due to dilution in the larger volumes of fat, serum, liver, and muscle present in obese children [20]. According to a previous Belgian study, vitamin D levels are significantly influenced by body composition [21]. Similarly, an Italian study showed that children who were overweight had higher levels of vitamin D deficiency than those with optimal weight [22]. Unfortunately, we did not have data on body composition available for our study.

Thirdly, children with limited sun exposure are more likely to suffer from a vitamin D deficiency [5]. Although Italy (45.6–41.5°) is located below 50° geographical latitude, we detected the same high prevalence of vitamin D deficiency as detected in Antwerp, located at 51.2° latitude [17,23]. Sunlight is thought to be the most important source of vitamin D, but the same high prevalence of vitamin D deficiency in both countries might indicate that other factors, like differences in lifestyle (outdoor play, veiling, etc.), diet (including food-fortification policies, vegetarian or vegan diet, etc.), vitamin D supplementation, or/and socio-economic dissimilarities, are as important as sun exposure [23]. By achieving sufficient solar exposure, the debate on proper photoprotection (to protect children against sunburn, cancer and photoaging) and the importance of sunlight exposure to establish adequate vitamin D levels rises in the majority of healthcare professionals and parents [24]. The effect of sunscreen application on vitamin D synthesis remains a topic of debate [1,24]. However, it is likely that the combination of sunscreens, along with covered clothing, dark pigmentation of the skin, and shade-seeking behavior are more likely to compromise vitamin D status in children [24]. Our results could imply that children lack sufficient safe sun exposure. Moreover, a low socioeconomic status is a risk factor for vitamin D deficiency [25,26].

Potential interventions should be multifaced too [7]. Firstly, our results seem to suggest that the WHO recommendation (administer supplements to all infants <1 year) is either not well-communicated and/or followed-up by healthcare professionals (pediatricians, general practitioners, well baby clinics, etc.) or not well administered/followed by parents. Raising awareness through straightforward communication and parental education could improve vitamin D levels in neonates. However, infants absorb calcium independently of vitamin D [27,28], so the clinical consequences of low levels in infancy might be limited. Moreover, the recommendation in Belgium is to supplement 400 IU of vitamin D daily for all children aged up to the age of six [15]. Based on the high prevalence of hypovitaminosis in children aged over 6 in our study, we suggest that maintaining vitamin D supplementation beyond the age of 6 could reduce the risk for vitamin D deficiency at a later age.

Secondly, individualized administration of supplements in children based on the presence of risk factors for low vitamin D production and intake (e.g., obesity, malabsorption diseases, veiling, . . .) should always be considered in clinical practice.

Lastly, sun exposure is the major source of vitamin D [7]. Research shows that playing outside for at least 15 min per day with uncovered arms and legs between 10 am and 3 pm and in the months between May and October would be sufficient to have adequate vitamin D levels throughout winter for children with pale skin [22]. The importance of extending sun-safe outdoor time in childhood should be highlighted properly in schools by promoting playing outdoor during all seasons.

Lastly, besides recommending supplements and adequate sun exposure, along with photoprotection, making vitamin D supplements and/or wider food fortification with vitamin D could also help to meet the vitamin D requirements in children [7,29,30]. Based on our study, action is needed to achieve sufficient vitamin D levels, especially in adolescents.

This study should be interpreted within the context of its limitations. Firstly, this is an observational retrospective study that does not consider biometrics, sun exposure, supplementation, nutrition, or ethnicity. Due to the nature of our study, detailed information was not available. Additionally, the indication for the blood test was not available, and this potential bias should be considered when interpreting the results. Secondly, children visiting a hospital were used to select our study population. This population does not perfectly reflect the general population. Well-known risk factors (chronic kidney disease, chronic liver disease, use of certain drugs, malabsorption, cholestasis, etc.) were not accounted for. However, given the large sample size, we consider the study population to be a representative sample, keeping in mind that unwell children or children with risk factors are more likely to be sampled when compared to their 'healthy' counterparts. Also, clinicians ordering the blood test could have a clinical suspicion of vitamin D deficiency that was not accounted for in our study, and would therefore create an additional bias. Lastly, because of the absence of data regarding PTH, phosphate, and bone density, conclusions about bone strength and clinical consequences in children with vitamin D deficiency cannot be drawn.

Additional research is required to investigate the consequences of our findings. Identifying causes might be effective to prevent and treat vitamin D deficiency. Also, research on specific approaches is needed to achieve normal vitamin D levels in children aged 0 to 18 years, especially in adolescents.

5. Conclusions

We report a very high prevalence of low vitamin D levels during childhood, mostly during late childhood and adolescence, in a large retrospective analysis of laboratory data in children. A huge proportion (over 70%) of these children do not meet the 30 ng/mL sufficiency threshold, especially children above 7 years of age. Additionally, we confirmed a seasonal variation with the lowest vitamin D serum values in winter and spring. No association between vitamin D and gender was found. It is essential that national health authorities are aware of the high prevalence of vitamin D deficiency in children, as the current Belgian recommendation would imply ceasing vitamin D supplementation at the age of six.

Author Contributions: Conceptualization, J.T. and A.R.; methodology, L.V.d.W., Y.V., J.T. and A.R.; software, L.V.d.W. and A.R.; validation, L.V.d.W., Y.V., J.T. and A.R.; formal analysis, L.V.d.W.; investigation, L.V.d.W., Y.V., J.T. and A.R.; resources, L.V.d.W., Y.V., J.T. and A.R.; data curation, L.V.d.W. and A.R.; writing—original draft preparation, L.V.d.W., J.T. and A.R.; writing—review and editing, L.V.d.W., Y.V., J.T. and A.R.; visualization, L.V.d.W., Y.V., J.T. and A.R.; supervision, Y.V., J.T. and A.R.; project administration, L.V.d.W., Y.V., J.T. and A.R. All authors have read and agreed to the published version of the manuscript.

Funding: This research received no external funding.

Institutional Review Board Statement: This retrospective regional study was conducted in accordance with the Helsinki Declaration for investigation in human subjects. The study protocol was approved by the Ethics Committee of the ZNA Hospitals Antwerp, Belgium (approval number N5514, date 8 December 2021).

Informed Consent Statement: Consent was waived because all data were pseudo-anonymously retrieved.

Data Availability Statement: Data are available upon reasonable request.

Acknowledgments: The authors thank Marian Piqueur, Wim Uyttenbroeck and colleagues for sharing the laboratory data and knowledge regarding the vitamin D assay used and Myriam Van Winckel for her valuable comments on the paper.

Conflicts of Interest: The authors declare no conflicts of interest.

References

1. Holick, M.F. The vitamin D deficiency pandemic: Approaches for diagnosis, treatment and prevention. *Rev. Endocr. Metab. Disord.* **2017**, *18*, 153–165. [CrossRef]
2. Antonucci, R.; Locci, C.; Clemente, M.G.; Chicconi, E.; Antonucci, L. Vitamin D deficiency in childhood: Old lessons and current challenges. *J. Pediatr. Endocrinol. Metab.* **2018**, *31*, 247–260. [CrossRef]
3. Gröber, U.; Spitz, J.; Reichrath, J.; Kisters, K.; Holick, M.F. Vitamin D: Update 2013: From rickets prophylaxis to general preventive healthcare. *Derm. Endocrinol.* **2013**, *5*, 331–347. [CrossRef]
4. Wacker, M.; Holick, M.F. Vitamin D—Effects on skeletal and extraskeletal health and the need for supplementation. *Nutrients* **2013**, *5*, 111–148. [CrossRef]
5. Lawson, D.E.; Paul, A.A.; Black, A.E.; Cole, T.J.; Mandal, A.R.; Davie, M. Relative contributions of diet and sunlight to vitamin D state in the elderly. *Br. Med. J.* **1979**, *2*, 303–305. [CrossRef]
6. Vissing Landgrebe, A.; Asp Vonsild Lund, M.; Lausten-Thomsen, U.; Frithioff-Bøjsøe, C.; Esmann Fonvig, C.; Lind Plesner, J.; Aas Holm, L.; Jespersen, T.; Hansen, T.; Christian Holm, J. Population-based pediatric reference values for serum parathyroid hormone, vitamin D, calcium, and phosphate in Danish/North-European white children and adolescents. *Clin. Chim. Acta* **2021**, *523*, 483–490. [CrossRef]

7. Saggese, G.; Vierucci, F.; Prodam, F.; Cardinale, F.; Cetin, I.; Chiappini, E.; De' Angelis, G.L.; Massari, M.; Miraglia Del Giudice, E.; Miraglia Del Giudice, M.; et al. Vitamin D in pediatric age: Consensus of the Italian Pediatric Society and the Italian Society of Preventive and Social Pediatrics, jointly with the Italian Federation of Pediatricians. *Ital. J. Pediatr.* **2018**, *44*, 51. [CrossRef]
8. Wharton, B.; Bishop, N. Rickets. *Lancet* **2003**, *362*, 1389–1400. [CrossRef]
9. Shroff, R.; Wan, M.; Nagler, E.V.; Bakkaloglu, S.; Fischer, D.C.; Bishop, N.; Cozzolino, M.; Bacchetta, J.; Edefonti, A.; Stefanidis, C.J.; et al. Clinical practice recommendations for native vitamin D therapy in children with chronic kidney disease Stages 2–5 and on dialysis. *Nephrol. Dial. Transpl.* **2017**, *32*, 1098–1113. [CrossRef]
10. Chapuy, M.C.; Preziosi, P.; Maamer, M.; Arnaud, S.; Galan, P.; Hercberg, S.; Meunier, P.J. Prevalence of vitamin D insufficiency in an adult normal population. *Osteoporos. Int.* **1997**, *7*, 439–443. [CrossRef]
11. Gentile, C.; Chiarelli, F. Rickets in Children: An Update. *Biomedicines* **2021**, *9*, 738. [CrossRef]
12. Ozkan, B. Nutritional rickets. *J. Clin. Res. Pediatr. Endocrinol.* **2010**, *2*, 137–143. [CrossRef]
13. Ao, T.; Kikuta, J.; Ishii, M. The Effects of Vitamin D on Immune System and Inflammatory Diseases. *Biomolecules* **2021**, *11*, 1624. [CrossRef]
14. Ladang, A.; Rousselle, O.; Huyghebaert, L.; Bekaert, A.C.; Kovacs, S.; Le Goff, C.; Cavalier, E. Parathormone, bone alkaline phosphatase and 25-hydroxyvitamin D status in a large cohort of 1200 children and teenagers. *Acta Clin. Belg.* **2022**, *77*, 4–9. [CrossRef]
15. De Ronne, N.; De Schepper, J. Vitamine D suppletie bij de zuigeling en het jonge kind: Aanbevelingen Kind en Gezin en Voedingscel Vlaamse Vereniging voor kindergeneeskunde. *Farm. Tijdschr. Voor Belg.* **2013**, *91*, 32–43.
16. Rabenberg, M.; Scheidt-Nave, C.; Busch, M.A.; Thamm, M.; Rieckmann, N.; Durazo-Arvizu, R.A.; Dowling, K.G.; Škrabáková, Z.; Cashman, K.D.; Sempos, C.T.; et al. Implications of standardization of serum 25-hydroxyvitamin D data for the evaluation of vitamin D status in Germany, including a temporal analysis. *BMC Public Health* **2018**, *18*, 845. [CrossRef]
17. Vierucci, F.; Del Pistoia, M.; Fanos, M.; Erba, P.; Saggese, G. Prevalence of hypovitaminosis D and predictors of vitamin D status in Italian healthy adolescents. *Ital. J. Pediatr.* **2014**, *40*, 54. [CrossRef]
18. Huybrechts, I.; Lin, Y.; De Keyzer, W.; Sioen, I.; Mouratidou, T.; Moreno, L.A.; Slimani, N.; Jenab, M.; Vandevijvere, S.; De Backer, G.; et al. Dietary sources and sociodemographic and economic factors affecting vitamin D and calcium intakes in Flemish preschoolers. *Eur. J. Clin. Nutr.* **2011**, *65*, 1039–1047. [CrossRef]
19. Webb, A.R.; Kift, R.; Durkin, M.T.; O'Brien, S.J.; Vail, A.; Berry, J.L.; Rhodes, L.E. The role of sunlight exposure in determining the vitamin D status of the U.K. white adult population. *Br. J. Dermatol.* **2010**, *163*, 1050–1055. [CrossRef]
20. Vranić, L.; Mikolašević, I.; Milić, S. Vitamin D Deficiency: Consequence or Cause of Obesity? *Medicina* **2019**, *55*, 541. [CrossRef]
21. Sioen, I.; Mouratidou, T.; Kaufman, J.M.; Bammann, K.; Michels, N.; Pigeot, I.; Vanaelst, B.; Vyncke, K.; De Henauw, S. Determinants of vitamin D status in young children: Results from the Belgian arm of the IDEFICS (Identification and Prevention of Dietary- and Lifestyle-Induced Health Effects in Children and Infants) Study. *Public Health Nutr.* **2012**, *15*, 1093–1099. [CrossRef]
22. Vierucci, F.; Del Pistoia, M.; Fanos, M.; Gori, M.; Carlone, G.; Erba, P.; Massimetti, G.; Federico, G.; Saggese, G. Vitamin D status and predictors of hypovitaminosis D in Italian children and adolescents: A cross-sectional study. *Eur. J. Pediatr.* **2013**, *172*, 1607–1617. [CrossRef]
23. Oskarsson, V.; Eliasson, M.; Salomaa, V.; Reinikainen, J.; Männistö, S.; Palmieri, L.; Donfrancesco, C.; Sans, S.; Costanzo, S.; de Gaetano, G.; et al. Influence of geographical latitude on vitamin D status: Cross-sectional results from the BiomarCaRE consortium. *Br. J. Nutr.* **2022**, *128*, 2208–2218. [CrossRef]
24. Passeron, T.; Bouillon, R.; Callender, V.; Cestari, T.; Diepgen, T.L.; Green, A.C.; van der Pols, J.C.; Bernard, B.A.; Ly, F.; Bernerd, F.; et al. Sunscreen photoprotection and vitamin D status. *Br. J. Dermatol.* **2019**, *181*, 916–931. [CrossRef]
25. Hoge, A.; Donneau, A.F.; Streel, S.; Kolh, P.; Chapelle, J.P.; Albert, A.; Cavalier, E.; Guillaume, M. Vitamin D deficiency is common among adults in Wallonia (Belgium, 51°30′ North): Findings from the Nutrition, Environment and Cardio-Vascular Health study. *Nutr. Res.* **2015**, *35*, 716–725. [CrossRef]
26. Raaijmakers, A.; Van Winckel, M.; Plaete, J.; Bovijn, L.; Van Overmeire, B.; Vandenplas, Y.; Stevens, G. Vitamine D voor kinderen in Vlaanderen. *Tijdschr. Voor Geneeskd. En. Gezondheidszorg* **2023**, *5*, 10–47671. [CrossRef]
27. Bass, J.K.; Chan, G.M. Calcium nutrition and metabolism during infancy. *Nutrition* **2006**, *22*, 1057–1066. [CrossRef] [PubMed]
28. Abrams, S.A. Vitamin D in Preterm and Full-Term Infants. *Ann. Nutr. Metab.* **2020**, *76* (Suppl. S2), 6–14. [CrossRef] [PubMed]
29. Moyersoen, I.; Lachat, C.; Cuypers, K.; Ridder, K.; Devleesschauwer, B.; Tafforeau, J.; Vandevijvere, S.; Vansteenland, M.; De Meulenaer, B.; Van Camp, J.; et al. Do Current Fortification and Supplementation Programs Assure Adequate Intake of Fat-Soluble Vitamins in Belgian Infants, Toddlers, Pregnant Women, and Lactating Women? *Nutrients* **2018**, *10*, 223. [CrossRef] [PubMed]
30. Moyersoen, I.; Devleesschauwer, B.; Dekkers, A.; Verkaik-Kloosterman, J.; De Ridder, K.; Vandevijvere, S.; Tafforeau, J.; Van Oyen, H.; Lachat, C.; Van Camp, J. A Novel Approach to Optimize Vitamin D Intake in Belgium through Fortification Based on Representative Food Consumption Data. *J. Nutr.* **2019**, *149*, 1852–1862. [CrossRef] [PubMed]

Disclaimer/Publisher's Note: The statements, opinions and data contained in all publications are solely those of the individual author(s) and contributor(s) and not of MDPI and/or the editor(s). MDPI and/or the editor(s) disclaim responsibility for any injury to people or property resulting from any ideas, methods, instructions or products referred to in the content.

nutrients

MDPI

Review

Targeting the Hallmarks of Aging with Vitamin D: Starting to Decode the Myth

Carmelinda Ruggiero [1,*], Laura Tafaro [2], Luisella Cianferotti [3], Flavia Tramontana [4], Ilaria Giovanna Macchione [1], Carla Caffarelli [5], Agostino Virdis [6], Marika Ferracci [1], Giuseppe Rinonapoli [7], Patrizia Mecocci [1,8], Nicola Napoli [9] and Valeria Calsolaro [6]

[1] Geriatric and Orthogeriatric Units, Division Gerontology and Geriatrics, Department of Medicine and Surgery, University of Perugia, 06156 Perugia, Italy; ilaria.macchione@ospedale.perugia.it (I.G.M.); marika.ferracci@ospedale.perugia.it (M.F.); patrizia.mecocci@unipg.it (P.M.)
[2] Division of Internal Medicine, Department of Molecular Medicine, Sapienza University of Rome, 00185 Rome, Italy; laura.tafaro@uniroma1.it
[3] Bone Metabolic Diseases Unit, Department of Experimental, Clinical and Biomedical Sciences "Mario Serio", University of Florence, University Hospital Careggi, 50134 Florence, Italy; luisella.cianferotti@unifi.it
[4] Unit of Endocrinology and Diabetes, Department of Medicine, Campus Bio-Medico University of Rome, 00128 Rome, Italy; f.tramontana@unicampus.it
[5] Division of Internal Medicine, Department of Medicine, Surgery and Neuroscience, University of Siena, 53100 Siena, Italy; carla.caffarelli@unisi.it
[6] Geriatrics Unit, Department of Clinical & Experimental Medicine, University Hospital of Pisa, 56124 Pisa, Italy; agostino.virdis@unipi.it (A.V.); valina82@gmail.com (V.C.)
[7] Orthopaedics and Traumatology Department, University of Perugia, 06129 Perugia, Italy; giuseppe.rinonapoli@unipg.it
[8] Division of Clinical Geriatrics, Department of Neurobiology, Care Sciences and Society, Karolinska Institutet, 17177 Stockholm, Sweden
[9] Unit of Endocrinology and Diabetes, Department of Medicine, Fundation Campus Bio-Medico University, 00128 Rome, Italy; n.napoli@unicampus.it
* Correspondence: carmelinda.ruggiero@unipg.it; Tel.: +39-075-578-3525 or +39-075-578-3839; Fax: +39-075-578-3878

Abstract: Aging is the result of several complex and multifactorial processes, where several agents contribute to an increased intrinsic vulnerability and susceptibility to age-related diseases. The hallmarks of aging are a set of biological mechanisms that are finely regulated and strictly interconnected, initiating or contributing to biological changes and anticipating several age-related diseases. The complex network of cellular and intercellular connections between the hallmarks might represent a possible target for the research of agents with pleiotropic effects. Vitamin D (VitD) is known to have a positive impact not only on muscle and bone health but also on several extra-skeletal districts, due to the widespread presence of Vitamin D Receptors (VDRs). VitD and VDR could be molecules potentially targeting the hallmarks of the aging network. To date, evidence about the potential effects of VitD on the hallmarks of aging is scarce in humans and mainly based on preclinical models. Although underpowered and heterogeneous, in-human studies seem to confirm the modulatory effect of VitD on some hallmarks of aging and diseases. However, more investigations are needed to clarify the pleiotropic effects of VitD and its impact on the hallmark of aging, hopefully highlighting the courses for translational applications and potential clinical conclusions.

Keywords: aging; vitamin D; hallmarks of aging; clinical studies

1. Introduction

Aging is a complex and multifactorial process, marked by a progressive decline in organ functions and an increased risk of age-related diseases and mortality [1,2]. At the biological level, it is a universal process characterized by accumulating damages linked to different mechanisms and pathways defined as "hallmarks" of aging [3]. These hallmarks manifest not merely during physiological aging; they exhibit tight connections with age-related changes, often anticipating alterations associated with age-related diseases and clinical phenotypes years in advance. Therefore, a better understanding of the features and mechanisms related to the hallmarks of aging becomes crucial for the early identification, prediction, and modulation of trajectories of organs and systems' age-related structural and functional changes. Moreover, experimental manipulation of hallmarks aligns with exacerbating or reducing aging processes and disease-related pathological changes.

To date, mammalians are characterized by twelve biological hallmarks of aging, namely genomic instability, telomere attrition, epigenetic alterations, loss of proteostasis, disabled macroautophagy, deregulated nutrient sensing, mitochondrial dysfunction, cellular senescence, stem cell exhaustion, altered intercellular communication, chronic inflammation, and dysbiosis [2]. These are biomolecular, time-dependent manifestations of alterations and mechanisms accompanying the aging process, enabling aging acceleration experiments and possibly deceleration interventions [4].

While these hallmarks have been extensively discussed in the following section, it is essential to recognize that they do not act as a single independent cause. They are incredibly complex and highly interrelated, complementing others in explaining some or all features of the aging and age-related disease processes [5]. Individually or synergistically, hallmarks can lead to molecular and cellular damage due to primary, antagonistic, and integrative mechanisms. The cumulative effects of the damages are universally adverse and might contribute to the initiation and progression of age-related diseases.

Understanding the mechanistic relationship among the hallmarks and identifying interventions that modulate such processes hold promise for developing strategies to intervene in human aging at the earlier onset of such changes and thus prevent age-related diseases.

Vitamin D (VitD) is renowned for its positive impact on musculoskeletal structure and functions, particularly in preventing falls, fragility fractures, impaired locomotion, and mobility disability among older adults with deficient plasma levels [6]. In the aging process, the decline in physical performance is recognized as the first phenotypical clinical feature of accelerated aging, preceding and contributing to the onset of muscle-skeletal chronic diseases and the burden of disability [7–9]. Consistently, aging is associated with a higher likelihood of vitamin D deficiency and dysregulated vitamin D function due to reduced sensitivity to 1,25(OH)2D3. This results from the combined effect of age-related decline in (Vitamin D Receptor) VDR expression, impairment of cutaneous vitamin D synthesis, and altered expression of vitamin D metabolic enzymes. In addition, poor nutrition and reduced exposure to sunlight are commonly seen in long-term care recipients—a population at a particularly high risk of deficiency [10]. Therefore, the clinical practice guidelines state that serum VitD levels of less than 25 nmol/L should be considered as vitamin D deficiency (hypovitaminosis D), 25–50 nmol/L as insufficiency, and over 50 nmol/L as sufficient [11]. Beyond its skeletal benefits, extensive preclinical research and some clinical studies support the extra-skeletal advantages of VitD at multi-organ levels through the 1α-hydroxylase enzyme, which is present in many tissues besides bone, including endothelial cells, vascular smooth muscle cells, cardiomyocytes, pericytes, neural stem cells, neurons, astrocytes, microglia, fibroblasts, osteoblasts, epithelial cells adipocytes, myocytes, monocytes, and macrophages.

The endocrine, intracrine, and paracrine actions of VitD may suggest its involvement in attenuating the aging process and the age-related pathological changes associated with multiorgan diseases [12]. Some studies suggest a role for VitD in influencing biological age at the epigenetic level since individuals with low VitD levels are biologically older than people with adequate levels [13]. Furthermore, VitD has been extensively investigated concerning age-related diseases, particularly cardiovascular disorders. Evidence links low levels of VitD to an increased risk of coronary artery disease, myocardial infarction, and heart failure. VitD supplementation has shown protective effects against the development and progression of cardiovascular diseases [14]. Therefore, randomized clinical trials were conducted to test VitD supplementation for preventing or improving non-skeletal diseases related to poor VitD status. Although the results were mixed, they support the multitargeting actions of low levels of VitD, which significantly affect the aging process by regulating cell homeostasis, counteracting oxidative and inflammatory damage, and cellular senescence.

This narrative review aims to summarize the available evidence regarding the effects of VitD on modulating age-related changes, either leading to or preventing diseases and frailty. Ultimately, we seek to determine whether VitD may be a feasible intervention for attenuating the hallmarks of aging, ultimately suggesting a role in promoting healthy aging or decelerating age-related changes anticipating multi-organ diseases.

2. Methods

To serve the purpose of this manuscript, we conducted Pubmed and Embase research, using keywords comprehending "VitD" and each hallmark of aging as previously described, including "genomic instability", "telomere attrition", "epigenetic alterations", "loss of proteostasis", "disabled macroautophagy", "deregulated nutrient sensing", "mitochondrial dysfunction", "cellular senescence", "stem cell exhaustion", "altered intercellular communication", "chronic inflammation", and "dysbiosis", by restricting the search to interventions conducted in the human species. We included English literature for the last ten years (the last search was performed in January 2024), with some exceptions for literature about general concepts.

3. The Hallmarks of Aging

Ten years after the initial report, in 2023, Lopez-Otìn et al. revisited the hallmarks of aging, expanding the list of common denominators of aging [2]. The hallmarks, identified across various organisms with special emphasis on mammalians, meet three main criteria. Firstly, they are time-dependent molecular or cellular manifestations of changes or alterations accompanying aging. Secondly, their experimental modulation may accelerate aging processes, and the therapeutic interventions targeting the hallmarks have the potential to decelerate, halt, or reverse age-related processes. The hallmarks of aging are currently categorized as primary, antagonistic, and integrative. Primary hallmarks are the causal agents of cellular damage; they encompass genomic instability, telomere attrition, epigenetic alterations, and loss of proteostasis. Antagonistic hallmarks include deregulated nutrient sensing, mitochondrial dysfunction, and cellular senescence. These initially counteract damages caused by primary hallmarks; however, incidentally or with prolonged pathway stimulation, the antagonistic mechanisms become harmful, contributing to accumulated deleterious age-related changes and processes. Integrative hallmarks, such as stem cell exhaustion, altered intercellular communication, chronic inflammation, and dysbiosis, are interconnected with other damage mechanisms and collectively contribute to functional age-related decline (Figure 1).

The hallmarks of aging are strongly related to each other, exhibiting intricate interactions, with each contributing to and exacerbating the effects of others. Genomic instability crosstalks to epigenetic alterations (e.g., through the loss-of-function mutation of epigenetic modifiers such as TET2), loss of proteostasis (e.g., due to the production of mutated, misfolded proteins), disabled macroautophagy (e.g., through the capacity of autophagy to remove supernumerary centrosomes, extranuclear chromatin, and cytosolic DNA), deregulated nutrient-sensing (e.g., because SIRT6 is an NAD+ sensor involved in DNA repair but also responding to nutrient scarcity), mitochondrial dysfunction (e.g., due to the mutation of mtDNA), cellular senescence (e.g., because DNA damage triggers senescence), altered intercellular communication (e.g., by hampering activation of communication pathways), chronic inflammation (e.g., because CHIP and leakage of DNA into the cytosol induce inflammation), and dysbiosis (e.g., because mutations in intestinal cells favors dysbiosis, whereas specific microbial proteins and metabolites act as mutagens). In addition, senescence can initiate inflammation, which in turn may worsen mitochondrial dysfunction and genomic instability. Epigenetic alterations, on the other hand, can influence gene expression, including the upregulation of proteins like p16 and p21, which promote senescence. Again, the loss of proteostasis, leading to the accumulation of misfolded and damaged proteins, triggers senescence pathways in response to stress. Then, dysregulation of the mTOR pathway, which regulates nutrient sensing, can increase mitochondrial dysfunction and impact on its biogenesis and function. Ultimately, increased ROS release associated with mitochondrial dysfunction can induce epigenetic alterations through DNA methylation and histone modification, which in turn creates a feedback loop of declining mitochondrial function [15]. The interdependence of the hallmarks unveils the complexity of the aging processes and the intricate links between aging and the organ's thresholds for diseases, and then highlights the need for conceiving aging as a whole and looking for organisms' macro-biomarkers that are able to quantify the energetic efforts associated with underlying regulatory or compensatory mechanisms leading to diseases [16]. Indeed, whether and how one specific hallmark's experimental accentuation or attenuation affects other hallmarks does not reveal molecular or cellular beneficial or detrimental effects immediately, given the onset of antagonistic or compensatory mechanisms that, over time, fail, given their intrinsic poor sustainability as disease markers. In addition, evidence is needed to identify whether aging is driven by the deterioration of cellular hallmarks converging on a common aging pathway in all cells or if independent mechanisms will result in different aging pathways in individual cells or tissues. A better understanding of the independence or interdependence of the hallmarks of aging will provide a rational basis for the development of novel therapeutics or interventions to modify the rate of aging.

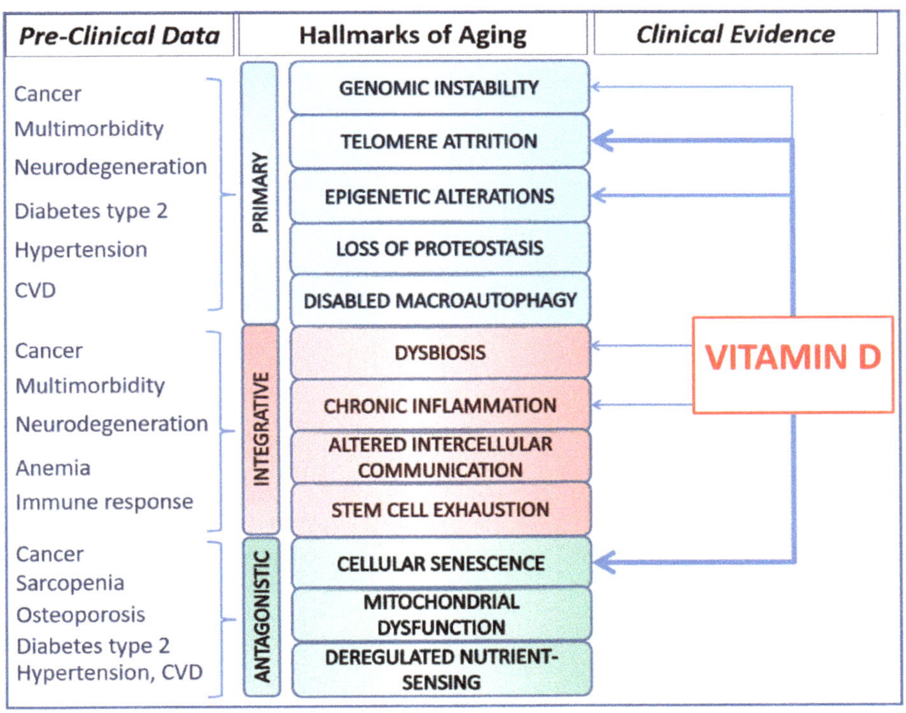

Figure 1. Pre-clinical data supporting the relationship between diseases and hallmarks of aging on the left, and available clinical evidence about the potential impact of Vitamin D on the hallmarks of aging, on the right. Legend: the thickness of the arrows is representative of the amount of available evidence.

3.1. Genomic Instability

The genomic instability pathway involves accumulating genetic damages caused by exogenous (chemical, physical, and biological) agents and endogenous challenges (DNA replication errors, chromosome segregation defects, reactive oxygen species (ROS), and spontaneous hydrolytic reactions). Such an accumulation of damage causes mosaicism, explaining the coexistence of normal and pathological aging. DNA lesions, such as mutations, deletions, translocations, telomere shortening, single- and double-strand breaks, chromosomal rearrangements, defects in nuclear architecture, and gene disruption caused by the integration of viruses or transposons, impact genes and transcriptional pathways, resulting in dysfunctional cells that compromise tissue and organism homeostasis [2]. When alterations impact stem cells, leading to their exhaustion or hampering their role in tissue renewal, the damage becomes relevant, accelerating the aging process and increasing susceptibility to age-related diseases [17]. DNA repair deficiencies have been linked to aging, suggesting that interventions reducing the mutational nuclear or mitochondrial load and enhancing or rerouting repair mechanisms may slow aging and delay the onset of age-related diseases. VitD emerges as a potential intervention in this context [18].

3.2. Telomere Attrition

Telomeres are nucleotide sequences that protect chromosome ends and preserve their genomic stability. Telomeres, together with the sheltering complex, ensure the correct segregation of genetic material during the cell division cycle and prevent DNA repair systems [19]. Due to the inability of the replicative DNA polymerases to complete the copy of telomere regions during cellular division cycles, a shortening of the telomere cap occurs (i.e., telomere attrition) with advancing age, then activating a DNA damage response cascade leading to cellular senescence or apoptosis [20]. Telomerase activity is affected with advancing age, and telomere shortening compromises cell function and life span and is associated with multiple age-related diseases, including cardiovascular disease, malignancies, dementia, osteosarcopenia, frailty, and other conditions [21]. Notably, deficient telomerase activity is also associated with pulmonary fibrosis, aplastic anemia, and dyskeratosis congenita, characterized by a hampered regenerative capacity of the affected tissues. Telomere attrition is adjustable to intrinsic and extrinsic factors, including diet and lifestyle, with evidence for protective effects due to VitD [22].

3.3. Epigenetic Alterations

Although the chromosomes carry genetic information, the epigenome is responsible for the functional use and stability of that valuable information, connecting the genotype with the phenotype [23]. Epigenetic alterations represent reversible, heritable mechanisms that occur without altering the underlying DNA sequence. Epigenetic changes can either be spontaneous or driven by external or internal influences, and they may potentially explain different patterns of aging between genetically identical individuals, such as identical twins [24]. A vast array of enzymatic systems is involved in the generation and maintenance of epigenetic patterns, including DNA methyltransferases, histone acetylases, deacetylases, methylases, and demethylases, as well as protein complexes implicated in chromatin remodeling or in ncRNA synthesis and maturation. Then, the major epigenetic alterations affecting aging include alterations in patterns of DNA methylation, abnormal posttranslational modification of histones, aberrant chromatin remodeling, and the deregulated function of non-coding RNAs [2]. Diet interventions, exercise, and pharmacological interventions based on epigenetic-related compounds, including VitD [13,25], are the most accepted strategies to target aging and age-related diseases in humans [14,24].

3.4. Loss of Proteostasis

The loss of proteostasis is the mechanism underlying the intracellular or extracellular accumulation of impaired proteins' aggregates (i.e., misfolded, oxidized, glycated, or ubiquitinylated proteins). Proteome integrity requires tight regulation and crosstalk among distinct proteostasis networks, from translation to degradation. Protein mutations increase their intrinsic tendency to misfolding and aggregation, hence saturating the protein repair, removal, and turnover mechanisms required to maintain a healthy state. In accelerated aging, the burden of misfolded proteins exceeds the capacity of cells to maintain proper proteome integrity and leads to disruptions of cellular function [1]. Over time, the enhanced production of erroneously translated misfolded or incomplete proteins, the slowed translation elongation, and the accumulation of oxidative damaged proteins sustain the collapse of the intracellular network for proteostasis. These alterations increasingly distract the chaperones from folding healthy proteins required for cellular fitness and then increase the failure of the quality control protein response to damage. Several preclinical models found evidence about the loss of proteostasis among neurodegenerative diseases. Dietary VitD supplementation has profound effects on protein homeostasis, with preclinical studies showing profound effects on IRE-1, XBP-1, and SKN-1 functions, suggesting the promotion of protein homeostasis and the slowing of aging processes [26].

3.5. Disabled Macroautophagy

Autophagy is an evolutionary, highly conserved, catabolic process that engulfs cytoplasmic materials and dysfunctional organelles by forming a double-layered autophagosome that fuses with the lysosome for the degradation and recycling of nutrients to maintain cellular homeostasis. Autophagy induces stress responses, and starvation, infection, and toxin exposure trigger the catabolic process to protect cells. Alterations in the autophagic functions compromise the intrinsic cellular housekeeping functions, limiting the recycling of metabolic substrates (i.e., free amino acids, carbohydrates, nucleotides, and lipids) for the biosynthetic process for cell maintenance and survival, favoring the accumulation of dysfunctional organelles and protein aggregates, and limiting the clearance of intracellular microbes through lysosomal degradation [27]. Indeed, autophagy primarily serves as a cytoprotective mechanism, and its impairment is associated with neural disorders, developmental abnormalities, inflammatory diseases, aging, and cancer [27]. VitD prompts cellular autophagic activities through genomic and non-genomic signaling, influencing various physiological functions along with calcium metabolism. Autophagic mechanisms regulated by VitD and its receptor exert protective effects in inhibiting oxidative stress and apoptosis, controlling cellular proliferation, differentiation, inflammation, and host immunity by activating antimicrobial defense mechanisms [27].

3.6. Deregulated Nutrient-Sensing

The nutrient-sensing pathways detect and respond to fluctuations in environmental nutrient levels. Physiologically, nutrient abundance activates anabolism and storage, whereas scarcity triggers homeostatic mechanisms that mobilize internal stores through autophagy. The nutrient-sensing network includes extracellular ligands (i.e., insulins, IGFs, and receptor tyrosine kinases) with which they interact and intracellular signaling cascades. These cascades involve the PI3K-AKT and Ras-MEK-ERK pathways and transcription factors, including FOXOs and E26 factors, which transactivate genes involved in diverse cellular processes.

Deregulated nutrient-sensing metabolic status is associated with a compromised somatotrophic axis, which includes the growth hormone (GH) and IGF-1, mainly produced by hepatocytes in response to the blood glucose concentration and activating intracellular signaling pathways. In preclinical models, diminished IGF-1 levels and GH/IGF-1 signaling are associated with improved stress defenses, autophagy, and cell survival via reduced PI3K/Akt and mTOR signaling. Moreover, GH receptor deficiency improves the defense from oxidative stress in healthy tissues and promotes apoptosis in neoplastic cells [28]. Modulating or perturbing nutrient-sensing node pathways seems to be a robust way to slow aging processes, extend the human lifespan, and protect against aging-related diseases. Ongoing studies testing caloric restriction (a reduction in caloric intake without apparent signs of malnutrition), pharmaceuticals, dietary patterns, or genetic interventions are waiting for more convincing results [29,30].

3.7. Mitochondrial Dysfunction

Mitochondrial structure and dynamics change with aging. Mitochondria from older subjects swell while their numbers decline and cannot replace themselves as quickly as in their optimal functional state [31,32]. Mitochondrial dysfunction follows the accumulation of mitochondrial DNA mutations, the increased production of ROS and related damage to cellular macromolecules, the impaired cellular bioenergetic and higher permeabilization of mitochondrial membranes, causing inflammation, altered stress responses, and cell death (when activators of caspases, nucleases, or other lethal enzymes are released from the intermembrane space). Notably, mitochondrial dynamics and quality control changes promote the accumulation of damaged mitochondria, which contributes to aging and age-related pathologies. In addition, several cardiac and neurodegenerative diseases, muscle atrophy, and sarcopenia recognize alterations in mitochondrial fission and fusion machinery [33]. Strategies aiming to counteract the age-related decline in L-carnitine levels, which may limit fatty acid oxidation by mitochondria, have revealed improvement or rescue of mitochondrial dynamics and quality control defects in a placebo-controlled trial on pre-frail subjects and older men [34]. Mitochondrial-derived peptides are emerging as key players in mitochondrial dysfunction through interactions with stress-sensitive transcription factors. Humanin is a microprotein encoded by mtDNA that declines with age, exhibits high levels in centenarians and their offspring, and emerges as a potential anti-aging factor linking organellar function to organismal homeostasis [35]. Humanin lacks contributions to cellular senescence, chronic inflammation, and cognitive decline [36]. Concerning mitochondrial dysfunction, VitD directly promotes and prevents protein oxidation, lipid peroxidation, and DNA damage and indirectly regulates autophagy, inflammation, epigenetic modifications, DNA abnormalities, and calcium and ROS signaling changes [37].

3.8. Cellular Senescence

Senescence is a state of permanent cell cycle arrest in response to different damaging stimula. During development, adulthood, and after injury, senescent pathways play a pivotal role in tissue remodeling, acting as a positive regulator of the regenerative potential and function of tissues. Senescent cells are also a strong safeguard against tumorigenesis, pacing irreversible proliferation arrest via cell and non-cell autonomous pathways. In aged organisms, the accumulation of cellular senescence and the release of secreted factors (SASP) adversely modify the behavior of senescent and non-senescent cells, remodel the extracellular environment, and sustain inflammation and tumorigenesis. Dysfunctional senescent cells may be powerful targets for slow-aging approaches, but the existence of beneficial senescence programs complicates the development of interventions without incurring toxicities [38]. Nutrition and a healthy lifestyle are intimately linked and can greatly impact cellular senescence. Nutritional elements can influence cellular senescence, with evidence for VitD from both in vitro and in vivo research [39]. In preclinical models, interventions targeting senescent cells have been shown to delay, prevent, or alleviate tissue damage and multiple disorders. Senolytic drugs may target selectively senescent cells, inducing their lysis, leading to promising strategies for preventing or treating multiple diseases and age-related conditions in humans. However, in some cases, their molecular targets have not been precisely identified and characterized, making them hard to introduce in clinical practice in the short term [40].

3.9. Stem Cell Exhaustion

Stem cell exhaustion refers to the numerical decline and impaired dynamics of the constitutional stem cells; both aspects are crucial strategies for living organisms. With unstable populations of proliferating stem cells, tissues and organs lose their ability to recover from primary or injury-induced damage and begin to fail. Cellular de-differentiation, plasticity, and cellular reprogramming, i.e., the plasticity of resident stem cells under normal conditions, are the main mechanisms for tissue renewal and repair. While stem and progenitor cells are all subject to the same hallmarks of aging, the mechanism of cellular reprogramming is thought to act autonomously on multiple cell types, with higher relevance for the long-term impact on tissue rejuvenation. Cellular re-programming consists of the conversion of adult somatic cells into embryonic pluripotent cells; transient reprogramming confers repair capacity to old tissues so that subsequent damage is repaired as efficiently as in young individuals; partial reprogramming encompasses features of the natural process of tissue repair, according to the finding that the epigenetic methylation clock accelerates soon after tissue injury and partially reverses during tissue repair [2]. A large body of the preclinical literature reports stem cell-based interventions' therapeutic impacts on tumors, fibrotic disorders, and tissue damage [41]. Human trials of any stem cell therapy must be approached cautiously due to the risk of genetic and epigenetic alterations occurring during their induction and manipulation. So far, insufficient clinical evidence supports the effectiveness and safety of anti-aging stem cell infusions [42]. Therefore, we are waiting for findings from trials looking for the effects of stem cell-based interventions on physical frailty and skin aging, which are still in their development phase [43].

3.10. Altered Intercellular Communication

Aging is characterized by altered intercellular communication due to extensive changes in the amount and secretion of cellular factors and hormonal signaling pathways. On one side, neural, neuroendocrine, and hormonal signaling pathways decline with aging due to intrinsic adaptations, including adrenergic, dopaminergic, and insulin/IGF1-based and renin-angiotensin systems and sex hormones. On the other side, intrinsic cellular alterations and environmental cellular exposures increase the SASP (senescent-associated secretory phenotype) and extracellular vesicles (EVs), acting as carriers for intercellular messaging through miRNAs that progressively impact the function of neighboring and distant cells, modifying the body's coordinated functions and responses. The intercellular communication derangements ultimately sum up the hallmarks of aging on its own bridging the cell-intrinsic hallmarks to meta-cellular hallmarks, including chronic inflammatory reactions, impaired immunosurveillance against pathogens and premalignant cells, and alterations in the bidirectional communication between the human genome and microbiome. The main focus in this area is exploring the effects of blood-borne systemic factors on pro-aging or pro-longevity properties. VitD is a fine-tuned immunomodulatory molecule that impacts immunosenescence and inflammaging, with evidence showing that adequate VitD levels help counteract the natural decline in immune surveillance [44].

3.11. Chronic Inflammation

Inflammation is a ubiquitous mechanism evolved to protect an organism from infection and injury. While the inflammatory cascade in response to acute infection or injury clears invading pathogens and incites wound healing, chronic inflammation is a potentially pathologic process arising from the perpetuity of the initial trigger or the dysregulation of signaling pathways that is harmful to health. The term "inflammaging" has been coined to describe such a low-grade, chronic inflammation process that progresses with age and impacts several processes. Chronic inflammation occurs because of multiple derangements stemming from the hallmarks of aging, and then it progressively increases according to different trajectories depending on the spatial–temporal degree of interaction between coexistent hallmarks and intrinsic and extrinsic factors [45]. The VitD immunomodulatory role is widely recognized to have positive effects on health span and lifespan. An adequate level of VitD counteracts inflammation with multilevel targeting effects, i.e., inhibiting the expression and signaling of TLR2, 4, and 9, reducing the production of cytokines such as TNF-a, IL-6, and Il-23, and repressing the activity of T cells recruiting chemokines [46]. The combination of VitD and curcumin supplements has been recently hypothesized to counteract neurodegeneration [44].

3.12. Dysbiosis

Dysbiosis refers to an imbalance in the gut microbiome, which can occur because of several factors related to host genetic variants (ethnicity), dietary factors, and lifestyle habits (culture), as well as environmental conditions (geography), which makes it challenging to unveil the relationships between the microbiota and pleiotropic age-associated disease manifestations. In the context of the human microbiome interaction, dysbiosis negatively impacts health, acting as a catalyst factor for fueling inflammation. Moreover, interventions aimed at restoring a youthful microbiome may extend health span and lifespan, with several inflammatory microbial metabolites, including SCFAs, and exopolysaccharides microbially derived, downregulating gut inflammation [47]. Caloric restriction diets induce structural changes in the gut microbiome, increasing the abundance of Lactobacillus and other species that influence healthy aging. A randomized, double-blind, placebo-controlled pilot study in overweight/obese insulin-resistant volunteers showed that the oral administration of pasteurized A. muciniphila improved insulin sensitivity and reduced insulinemia and plasma total cholesterol levels [48]. Notably, the composition of the intestinal microbiome could be altered by vitD deficiency, and supplementation has been determined to improve gut microbiomes with a positive impact on their overall health and immune system function. Ultimately, fecal microbiota transplantation opens the possibility of manipulating the gut microbiota with pre-, pro-, and post-biotics to rejuvenate the immune system and connected organs, prospecting the analysis of microbiome data for personalized patient care [49].

4. Vitamin D and Its Potential Impact on the Biological Hallmarks of Aging

VitD is a highly regulated fat-soluble hormone, and biosynthesis starts after subcutaneous UV irradiation of 7-dehydrocholesterol. UV radiation plays a crucial role in initiating the conversion of 7-dehydrocholesterol to pre-vitamin D3, which further transforms into cholecalciferol, following the mevalonate pathway and the Kandutsch–Russell branch of the sterol pathways. Interestingly, VitD shares common metabolic pathways with cholesterol; their biosynthesis starts with acetyl coenzyme A (AC-CoA) in the cytosol and recognizes the influence of similar SNPs [50]. Cholecalciferol then converts into calcidiol (25(OH)D) after liver metabolic activation and finally into the active form calcitriol (1,25(OH)2D). The active form of VitD is involved in several complex metabolic pathways, occurring in several cellular types, and spanning extracellular regions, cytosol, lysosome, endoplasmic reticulum, and mitochondria. Both circulating calcidiol and calcitriol bind to the VitD binding proteins (VitD-BP), then calcitriol dissociates in tissues to bind to the VDR, a nuclear receptor transcription factor, and calcidiol seems able to bind accessory pockets of the VDR [51]. The calcitriol uptake into cells occurs in both bound and unbound forms and regulates calcium and phosphorus homeostasis through binding to VDR [52]. Beyond bone metabolism, VitD-BPs have immunomodulatory functions, being a negative acute phase protein with levels dropping by about a third in patients with acute health conditions [53]. Notably, VDR seems essential for mitochondrial bio-energetic function through the regulation of key mitochondrial gene sets. The role of single nucleotide polymorphisms (SNPs) in the VDR gene has been largely investigated but yet makes definitive conclusions difficult due to variable methodologies and different demographics among study populations [54]. Additionally, no evidence exists proposing a molecular basis for the differences seen between those who carry VDR SNPs and those who do not; clarification of the mechanisms underlying the functional consequences of these SNPs would provide further information into the role of the vitamin D/VDR axis and the extent to which its genetic component contributes to our physiology [55].

Considering the wide range of functions that VitD exerts on many cellular systems, it would not be surprising to find an impact on the hallmark of aging. However, the literature about the potential role of VitD on hallmarks of aging is scarce, limited to a few clinical studies and mainly preclinical models, probably due to the complexity of pathways involved in the aging process. Nevertheless, the fact that VitD exerts genomic actions has been known for a few years now. VitD acts through binding to VDR, after which heterodimers with retinoid receptors are created, and interacts with VitD response elements (VDREs), enhancer elements in DNA. Indeed, classic VDR belongs to the nuclear receptors and directly affects gene expression. It could induce gene transcription and also suppress gene expression, depending on the VDR interactions [56]. In addition, 1,25-dihydroxyVitD may also act through a receptor in the cytoplasmic membrane associated with a special type of lipid rafts called caveolas. Through the VDR in the cytoplasmic membrane, 1,25-dihydroxyVitD can activate signaling pathways and elicit a rapid cellular response [57]. The wide gene targeted by VitD and VDRs justifies the pleiotropic functions of VDRs. Table 1 summarizes the main human studies investigating the effects of VitD supplementation on the main hallmarks of aging. A detailed description of the studies is reported and discussed in the following sections.

4.1. *Vitamin D and Genomic Instability*

Associations between VitD and markers of DNA integrity and stability have been mainly presented in studies conducted in human cultured cells and observational studies conducted in patients with comorbid diseases, such as Type 2 Diabetes Mellitus (T2DM), obesity, infertility, cancer patients, and the general population [2,58,59]. Oncogenic-induced senescence has been the drive to demonstrate VitD and VDR genomic functions. Without going deeply through oncogenic pathways, oncogene-induced senescence is the consequence of the expression of oncogenic RAS, as a tumor suppressor mechanism. In more detail, the expression of RAS leads to the down-regulation of VDR, which in turn determines a reduction in BRCA1 and 53BP1, which are the main actors involved in genome stability. The levels of those factors appear to improve with VitD supplementation [60]. In a placebo-controlled trial involving 92 patients with a history of colorectal adenoma, there was no clear evidence of a reduction in 8-hydroxy-2-deoxyguanosine, a marker of DNA damage in normal colorectal epithelial crypt cells. The daily administration of 800 IU VitD3 alone for six months led to a significant increase in Bax expression, an apoptosis promoter, without changes in Bcl-2 expression, an apoptosis inhibitor [61]. However, a randomized control trial conducted in 92 people with VitD insufficiency (<30 ng/mL) showed that VitD supplementation (2000 IU daily for 3 months) decreased the percentage of DNA damage and oxidative parameters when compared to the control group, either with or without diabetes [62].

A potential role of VitD in modulating genomic instability has been assessed in T2DM. Indeed, poor glucometabolic control might affect genomic stability; at the same time, VitD plays a role in sustaining DNA damage repair. In their study, Fagundes et al. aimed to evaluate the effect of VitD in improving glucose and lipid metabolism as well as the modulation of the genomic instability derived from T2DM. The Damage Index obtained from a comet essay was reduced during VitD supplementation therapy, together with an anti-mutagenic action. VitD supplementation led to a significant decrease in nitric oxide (NO) and total thiols and an increase in the concentration of reduced glutathione (GSH), leading to a decrease in oxidative processes in cells. Even though this result might be debated, it is important that the antioxidant function of VitD may be a protective mechanism against DNA damage [63].

4.2. *Vitamin D and Telomere Attrition*

Several human studies have aimed to elucidate the relationship between VitD and telomere biology. A positive correlation between serum 25(OH)D levels and telomere length was found in peripheral blood leukocytes among women in the twins UK cohort. Even after adjusting for various factors, the difference in telomere length between the highest and the lowest 25(OH)D tertiles equated to a 5-year difference in telomeric aging [64]. The Nurses' Health Study, conducted on 1424 women, confirmed a positive correlation between telomere length and serum 25(OH)D concentrations. However, no significant association was found with 1,25(OH)2D3 levels. Genetic factors related to VitD metabolism did not significantly affect telomere length [65]. However, the association between 25(OH)D or

1,25(OH)2D3 levels and peripheral blood leukocyte telomere length was not confirmed in 2483 men [66] and 5096 young adults from the Northen Finland Birth Cohort 1966 [67]. Nevertheless, the study identified an inverse association between body mass index (BMI) and telomere length, even after adjusting for various factors, including 25(OH)D levels [67].

A more recent analysis conducted in older adults ($\geq$85 years old) examined the association between 25(OH)D concentration and telomere length in peripheral blood mononuclear cells (PBMCs) over time. They found a significant positive association at baseline but inconsistent relationships at subsequent time points [68]. Focusing on patients with systemic lupus erythematosus and healthy controls with VitD deficiency, again a significant correlation appeared between telomere length and serum 25(OH)D concentrations [69]. Ultimately, active VitD treatment was associated with a greater telomere length in PBMCs in a small sample of hemodialysis patients, supporting its potential protective role [70]. The effect of an intervention based on the combination of regular-strength exercise and supplements that included branched-chain amino acids (BCAAs), calcium, and VitD was tested on the telomere length and the expression of telomeric repeat-containing RNA (TERRA). Such an intervention significantly increased TERRA but did not alter telomere shortening [71]. TERRA is central to the modulation of telomere length. In cells with longer telomeres, TERRA competes with the telomerase's DNA substrate or it enhances the catalytic reverse transcriptase subunit of the enzyme to inhibit the elongation of telomeric repeats [72]. On the contrary, in cells with shorter telomeres, TERRA promotes telomere lengthening by facilitating the recruitment of telomerase. It is possible that the findings reflect a lag in the increase in telomere size. The cross-sectional design of this study limits any causal associations between TERRA and telomere length, which need to be explored using a prospective study design. By using summary-level data from the largest genome-wide association studies, a Mendelian randomized study investigated the causal relationship between serum VitD (n = 73,699) and telomere length (n = 37,684). The findings did not support any causal effect of VitD on telomeric length [73].

Obesity is a condition characterized by low-grade chronic inflammation with increased oxidative stress status, which in turn negatively impacts telomerase wellness. It has already been demonstrated that obesity accelerates aging and is associated with shorter telomeres. Telomerase activity maintains telomere lengths and prevents T cell senescence, preserving long-term immune function. On this basis, Zhu et al. recently evaluated whether VitD supplementation might increase telomerase activity in a population of overweight African American subjects compared to the control group. Telomerase activity was measured using the Telomeric Repeat Amplification Protocol (TRAP). At baseline, PBMC telomerase activity did not differ between the two groups, but VitD supplementation significantly increased telomerase activity, persisting after an adjustment for sex, BMI, and age, supporting the idea of a beneficial effect of VitD on telomer wellness [74]. However, telomere length was not measured in this study [74].

4.3. Vitamin D and Epigenetic Alterations

On a different note, the concept of epigenetic alterations leading to accelerated aging has been developed in recent years and has been associated with several diseases, such as obesity, Werner syndrome, Huntington's disease, and Down syndrome. The cumulative effects of epigenetic alterations are measured as the DNAm (DNA metilation) and seem to be highly correlated with the individual's chronological age [75]. The possible role of VitD in modulating epigenetic regulation has been hypothesized, considering the association between global DNA hypometilation, leucocyte DNA metilation changes, and increased global metilation levels after VitD administration [76,77]. In a recent study conducted to evaluate the effect of VitD on epigenetic aging in a cohort of young, obese African women, DNAm age was correlated with chronological age in patients with low VitD levels, while the supplementation decreased DNA metilation and slowed epigenetic aging [75]. Pregnancy has been another condition of interest for the study of epigenetics and DNA methylation. As part of the Southampton Women's Survey (SWS) mother–offspring study, it has been demonstrated that methylation at the retinoid-X-receptor-alpha (RXRA) locus in the umbilical cord was associated with the offspring bone mass. The relevance of this discovery is due to the fact that RXRA is part of VitD intracellular signaling, and greater VitD levels were associated with a lower RXRA methylation burden [78]. From the cohort of the Maternal VitD Osteoporosis Study (MAVIDOS), Curtis et al. measured the DNAm at a specific site in the RXRA locus (10 CpG) and found that in the cohort of mothers receiving cholecalciferol, the methylation levels in the RXRA-specific site were significantly lower compared to the placebo group [79].

Anderson et al. implemented an epigenome-wide approach to identify differentially methylated loci among infants born to mothers supplemented with 400 IU (i.e., control) versus 3800 IU (i.e., intervention) daily of VitD. Among mothers, supplementation with 3800 IU of VitD per day was associated with mean 25(OH)D levels reflective of a sufficient maternal VitD status by birth among the intervention group, as well as significant gains in maternal leukocyte DNA methylation associated with genes involved in cell migration/motility, development, and growth. The implications of methylation gain for gene expression and pathway activity could be related to blood vessel development, which is central to the biological processes associated with the demands for a hypertrophic maternal circulatory system. The continued development of the circulatory needs of the placenta and maternal VitD status have been shown to affect placental vascular endothelial growth factor gene expression. Among infants, those born to mothers in the intervention group showed methylation gain at loci with biological implications for the developmental processes, including bone, lung, metabolic, and nervous systems. Among infants of mothers in the intervention group, methylation loss was associated with genes known to play a role in metabolic processes and signal transduction pathways, a finding also noted in examining the leukocytes of their mothers [80].

4.4. Vitamin D and Lack of Proteostasis

How VitD participates in maintaining proteostasis has been studied in the preclinical model of C. Elegans, in which DAF-16, HSF-1, and SKN-1 are associated with the premature accumulation of insoluble proteins and alterations of protein homeostasis. In their study, Mark et al. found that VitD feeding was able to suppress the toxicity of amyloid-beta and slow protein insolubility, reducing protein toxicity in favor of longevity [26]. Preclinical studies have shown a role of VitD in muscle proteostasis, with a fine regulation in place between muscle protein synthesis (MPS) and catabolism (muscle protein breakdown: MPB). The ubiquitin-proteasomal pathway (UPP) is one of the proteolytic systems where VitD plays a recognized role. In fact, the deficiency of VitD in rats leads to increased MPB and increased expression of ubiquitin conjugating enzyme and ubiquitin conjugates, with no increase in lysosomal enzymes; as a result, VitD deficiency seems to contribute to increased non-lysosomal proteolysis [81]. Moreover, the lack of VitD signaling leads to muscle atrophy of type II fibers [82], sustaining the critical role of VDR in muscle metabolic flexibility and pancreatic insulin response. Recent studies have suggested that VDR-deficient mice are predisposed to utilize fatty acids as the primary energy source and accordingly exhibit an increase in the oxidative type I and type IIa fibers and atrophy affecting both of these muscle fiber types [83]. On the other hand, VitD has a role in the phosphorylation of mitogen-activated protein kinase (MAPK) pathway, influencing cellular signaling involved in myogenesis [81]. Adenosine monophosphate-activated protein kinase (AMPK) represents another proteostasis regulator that is linked to VitD levels. Indeed, AMPK is activated during metabolic stress, acting as a nutrient sensor and positively regulating the cellular response to stressors. It is known that AMPK expression declines with age, but in vitro studies suggest that its activity seems to increase after treatment with 1,25(OH)2D3. Moreover, VitD-deficient rats had reduced activation of sirtuin-1, whose activity was derived from AMPK activation [55]. So far, evidence in this regard comes from preclinical models, but considering the widely recognized role of VitD supplementation for muscle health, some analogue mechanisms in humans could be hypothesized.

4.5. Vitamin D and Mitochondrial Dysfunction

Protein homeostasis and cellular respiration should not be considered separately. The finding of novel mitochondrial localization of VDR has shed light on the impact of VitD on mitochondrial function. Indeed, after binding VRD, VitD seems able, in preclinical models, to reduce mitochondrial respiration, consequently preventing the excess of ROS [84]. In aged mice, the expression of the protein Nrf2 (nuclear factor erythroid 2–related factor 2, a regulator of cellular defense from oxidants), which is known to improve mitochondrial function, is reduced, but increases after treatment with calcitriol, further confirming the role of VitD in mitochondrial homeostasis [85]. The regulation of mitochondrial function has been seen as helpful in reversing age-related hypertension in young male mice [85]. Again, due to the known role of mitochondria in neurodegeneration and Alzheimer's disease in particular, Mohanad et al. found that VitD was able to improve the oxidative stress and mitochondrial function in a preclinical model of induced AD, through calcium/calmodulin-dependent protein kinase kinase 2 (CAMKK2)-mediated phosphorylation of Sirtuin1 (SIRT1) [86].

One of the districts that highly depends on adequate energy production and cellular respiration is the muscular one. In chronic obstructive pulmonary disease (COPD), besides the pulmonary impairment, several extra pulmonary districts are involved. In particular, type I muscle fibers suffer from reduced airflow and oxygenation, and the muscular structure switches from slow to a fast-glycolytic metabolism. Thus, increased fatigue and reduced endurance lower extremity strength with consequent impaired mobility, and finally, sarcopenia or cachexia represent possible consequences [87]. VitD deficiency has been demonstrated to be linked to mitochondrial dysfunction, ATP depletion, increased production of ROS, with the known damage that oxidative stress causes to muscular tissue, and VitD supplementation, which has a positive impact on improving lung function, inspiratory muscle strength, and oxygen uptake. Ca^{2+} homeostasis is a key player in muscle health; moreover, Ca^{2+} represents, together with ROS, secondary messages in cellular signaling. VitD is the main regulator of Ca metabolism, and its deficiency may impair calcium uptake in mitochondria, compromising adequate homeostasis. Again, VitD plays a role in protein synthesis degradation and peroxidation, actively impacting the ATP-ubiquitin-dependent system and nitrosative stress [87]. Several other examples of evidence support the VitD beneficial effect on the attenuation of mitochondrial oxidative stress. In rats, VitD attenuated the oxidative stress neurotoxicity induced by cyanide. In rats' cardiovascular systems, VitD was proven efficient in reversing the oxidative cardiac injury induced by isoproterenol, reducing H_2O_2 levels, and increasing anti-oxidant mechanisms. Moreover, VitD modulates the expression of monoamine oxidase, known to be a mitochondrial pro-oxidative enzyme, and restores vascular function [88]. The VitD beneficial function is exerted through VDRs activation; in mouse hearts, VDR's activation was able to reduce tissue ischemia/reperfusion stress, and its overexpression ameliorated the myocardial infarct size and increased cardiac function.

Improvement in mitochondrial stress with VitD has also been seen in the central nervous system; in fact, pre-treatment with calcitriol showed attenuation of oxidative stress and ROS production in conditions of mild hyperomocysteinemia. The same protective effect was demonstrated in the heart tissue when exposed to the same condition. Interestingly, VitD combined with lipoic acid reduced mitochondrial stress in astrocytes and neurons, both activated by H_2O_2 and streptozotocin [88]. With aging, the accumulation of AGE leads to several alterations in the cardiovascular system, such as stiffness, atherosclerosis, plaque formation, and endothelial dysfunction. In cardiomyocyte cells treated with AGE, mitochondrial dysfunction was associated with contractile impairment. Treatment with VitD reduced AGE accumulation in rats, therefore suggesting a further potential protective role [89]. Mitochondrial dysfunction, and consequently, altered placental metabolism and increased ROS production, is a feature widely recognized in obese pregnant women; similarly, obese women seem to be more frequently VitD-deficient. Phillips et al. evaluated whether trophoblast cells treated with VitD could show an improvement in mitochondrial respiration: in vitro treatment with calcitriol showed an increased expression of VDR, both in obese and non-obese women, while no significant difference was found in oxygen respiration after treatment in non-obese women, and some improvement was seen in obese ones [90]. Crohn's disease (CD) represents a chronic invalidating disease, and intestinal fibrosis is a common complication. Epithelial cells switch to mesenchymal cells through a process called epithelial–mesenchymal transition (EMT). Aside from the knowledge that patients with CD have reduced VitD levels and that VDD could lead to intestinal fibrosis, Yu et al. investigated whether VDR might have a role in modulating intestinal fibrosis through mitochondrial regulation. In VDR-knockout mice, epithelial mitochondria showed altered morphology, accumulation of ECM in the interstices, and reduced functional genes, suggesting a potentially protective role of VDR activation against intestinal fibrosis [91]. Finally, sarcopenia represents a widely recognized clinical syndrome, characterized by impaired physical performance and related to poor outcomes. The role of VitD in sarcopenia has been evaluated since hypoVitD is linked to muscle weakness, reduced strength and mass, fatigue, and reduced mitochondrial activity. However, the results from intervention studies evaluating the role of VitD are conflicting and inconclusive, recommending further studies to better understand and define the possible role of VitD supplementation in sarcopenic patients [92].

4.6. Vitamin D and Cellular Senescence

There are several insults that might activate the cellular senescence process. When the cell reduces proliferative capacity, it remains in the Eg1-g2 PHASE, and then interrupts divisions and changes morphology. The hallmark of senescence is the senescence-associated secretory phenotype (SASP). Proinflammatory cytokines released by the secretome may increase and trigger secondary senescence in the surrounding environment. Again, VitD supplementation seems to be related to decreased cellular senescence and SASP [93]. Part of the aging process is related to the aging of the immune system due to thymic involution and function and the increased number of T immature T lymphocytes, with the immune phenotype changing towards a lower CD4/CD8 ratio and a decreasing amount of T CD8+CD28+ cells. The inevitable process of immunosenescence exposes older people to recurrent infections and cancers; moreover, chronic disease and low-grade inflammation represent a further step down in immune system efficiency [94].

A recent trial aimed to test the possible effect of a nutraceutical supplement (comprehensive of Sambucus nigra, zinc, tyndallized Lactobacillus acidophilus (HA122), arabinogalactans, VitD, vitamin E, vitamin C, and group B vitamins) on some specific signatures of immunosenescence in hospitalized older adults. These supplements determined a reduction in inflammatory mediators such as IL-6, CRP, and lymphocyte number, and they were also associated with perceived better wellness [55,95]. Considering the effect of VitD on the immune system, mainly exerted through the inhibition of monocyte and cytokine overproduction and direct action on T lymphocytes, Rizka et al. evaluated the potential effect of alphacalcidol supplementation in an inflammatory profile (Il-6, IL-10, TNF, CD4/CD8 ratio, CD8+, and CD28-) in 110 older fit or frail subjects. The population was randomized to alphacalcidol or placebo and administered orally for ninety days. The cohort receiving alphacalcidiol had a better immune profile across the whole cohort as well as the cohort divided according to frailty status [96]. These results support the immunomodulatory function of VitD and highlight the importance of maintaining sufficient levels of VitD, which seem crucial in immunosenescence and chronic conditions characterized by low-grade chronic inflammation. In addition, calcium and calcium signaling have been established as critical factors in the implementation and regulation of cellular senescence. Increased intracellular calcium accumulation occurs in different cell types and in response to many different senescence-inducing stimuli. In senescent cells, intracellular calcium concentration increases in the cytosol and mitochondria, coming from the extracellular space through plasma membrane channels, the release by the endoplasmic reticulum, and mitochondrial dysfunction. Calcium fluxes between different subcellular compartments, calcium channels, calcium-binding proteins, calcium-regulated enzymes, and transcription factors also play pivotal roles in promoting senescence. However, a thorough assessment of the functional contribution and regulation of each of the factors involved in calcium signaling is still needed to fully understand it [97].

4.7. Vitamin D and Stem Cell Exhaustion

The influence of VitD on stem cells (SCs) mainly refers to in vitro studies conducted on fetal or adult osteoblasts. The administration of VitD per se or loaded onto nanoparticles and other nanostructures is associated with great potential in bone regeneration and improved stem cells' differentiation into osteoblasts [98].

VDR signaling also plays a vital role in the modulation of neuronal SCs, which exhibit a prominent role in developing the entire nervous system during the embryonic stage; even in adult brains, neuronal SCs are located in restricted regions and support the replacement of damaged neurons [99]. VDR-mediated cellular signaling cascades are involved in enhancing neuronal SCs through Wnt/β-catenin and Sonic Hedgehog pathways. Based on the canonical signaling of Wnt in activating neurogenesis in the brain and maintaining balance between activated neural stem cells (aNSC)/qNSC avoiding depletion, it is crucial to maintain persistent VDR activity. Therefore, deficiency in VitD could cause an imbalance in Wnt signaling and inhibit the gradual neural SC stimulation required to maintain a neurogenic rate [100]. However, more research is needed in this field to identify potential VDR agonists targeting neuronal SCs. More recently, preclinical studies identified a potential role of VitD on SCs from patients with acute myeloid leukemia (AML). These patients usually display insufficient or deficient VitD levels, which are significantly associated with poorer overall survival, both at diagnosis and before allogeneic stem cell transplantation. There is seminal evidence from animal studies that VDR agonists may impair leukemic stem cell activity, while the lack of VDR results in increased numbers of hematopoietic and leukemia stem cells and quiescent hematopoietic stem cells, suggesting a function of VitD as regulators of stem cell homeostasis [101,102]. However, therapeutic interventions with VitD analogues have resulted in inconclusive data in humans due to adverse events related to systemic hypercalcemia [103]. Possibly more promising results could be derived from recent pilot studies evaluating the effects of adding VitD to the adipose tissue-derived stromal/stem cell (ASC) infusion in patients with Type 1 Diabetes [104].

4.8. Vitamin D and Altered Intercellular Communication

Very scarce literature is available regarding the possible role of VitD in intercellular communication, and it is mainly focused on bone cells. It is already known that fibroblasts and other osteoblast-lineage cells lead to the formation of osteoclasts. However, it is not clear whether intercellular interactions have an impact on the expression of genes implied in osteoclastogenesis. In their study, Bloemen et al. evaluated the effect of the cell–cell interaction on the micro-RNA (mRNA) expression of the molecules involved in osteoclast formation in cell cultures. Moreover, they evaluated the formation of tartrate-resistant acid phosphatase (TRACP)-positive cells as well as their ability to resorb bone. Interestingly, their data showed that osteoclastogenesis-related genes were up-regulated in co-cultures compared to mono-cultures and were overpowered by both VitD and dexamethasone [105]. Intercellular communication is also a contributing factor for receptor expression on osteoblast-like cells, which is fundamental for hormone and neurotransmitter binding and signaling. Several years ago, Schirrmacher analyzed the electric coupling in a gap junction of rat osteoblast-like cells after stimulation with VitD, 17β-estradiol (17β-E2), the neurotransmitter vasoactive intestinal peptide (VIP), and the excitatory amino acid glutamate (Glu). They found that the electric coupling response was cell-pair-specific when exposed to the above-mentioned hormones. Specifically, VitD increased the coupling in some cell pairs but not in others [106]. Considering the pleiotropic actions of VitD at cellular levels, it could possibly have an influence on intercellular communication as well, and more studies are warranted to better address this topic.

4.9. Vitamin D and Chronic Inflammation

VitD influences the innate and adaptative immune responses given to the wide distribution of VDR among immune cells and their intrinsic ability to convert VitD metabolic precursors into the active form, and then allows for an autonomic regulation of the active VitD concentration in an inflamed site [107]. Although several preclinical and epidemiological studies confirmed the inverse relationship between VitD and chronic inflammation, intervention studies addressing the impact of VitD supplementation on low-grade inflammation are not consistent. Indeed, VitD supplementation reduces the high-sensitivity C-reactive protein (hs-CRP) in patients with diabetes [108], psychiatric disorders [109], polycystic ovary syndrome [110], and advanced kidney disease [111], and reduces the tumor necrosis factor-α (TNF-α) in patients with diabetes [112]. However, a systematic review of meta-analyses and randomized controlled trials (RCTs) found that VitD3 supplementation at a range of doses had no significant effect on the biomarkers of systemic inflammation and hypothesized that a low VitD status is a consequence of ill health rather than its cause [113]. The VITAL study aimed at the primary prevention of cancer and CVDs and was conducted among 25,871 women $\geq$ 55 and men $\geq$ 50 years of age, with a mean treatment period of 5.3 years. It confirmed that VitD supplementation (2000 IU/daily), with or without n-3 FAs, decreased hs-CRP by 19% at year 2, although the reduction was attenuated at 4 years. Other inflammatory biomarkers (IL-6, IL-10, and TNF-α) were not significantly altered at year 2 or year 4, while n-3 FAs, with or without VitD3, did not significantly affect these biomarkers at either time point. Therefore, the authors partially confirmed a potential role of VitD supplementation in modulating the chronic inflammatory process, systemic inflammation, and possibly autoimmune disease progression [114].

4.10. Vitamin D and Dysbiosis

Dysbiosis and gut microbiota dysfunction may result from the interplay of several factors, and chronic inflammation represents a possible drive for them. HIV-1 infection is an example of a chronic disease where VitD seems to counteract dysbiosis. Indeed, patients with HIV-1 infection are characterized by persistent immune activation led by lipopolysaccharides (LPS) and microbial products migrating through a defective gut mucosal barrier. They often present with hypovitaminosis D, and, as a consequence, a concomitant dysregulation of the gut microbiota has been described. The immunomodulatory function of VitD seems to be exerted through the induction of the antimicrobial peptide LL-37, which is protective via the activation of autophagy, the inhibition of HIV-1 transcription, and pathogen killing [115]. Moreover, preclinical and clinical studies demonstrated a potential role of VitD in stabilizing the epithelial tight junction, as well as regulating the intestinal microbiota composition, reducing pro-inflammatory intestinal probacteria. Based on these premises, Missalidis et al. evaluated whether VitD or phenylbutyrate (PBA) may have modulatory action on microbiota composition, metabolite production, and immune activation. As a result, VitD and PBA supplements increase VitD levels, without having a significant improvement in the circulating antimicrobial peptide LL-37 and an actual impact on microbiota [115]. Another chronic inflammatory condition linked to gut microbiome dysbiosis is osteoarthritis (OA). Considering that microbiota dysfunction may lead to chronic low-grade systemic inflammation through LPS, and that LPS is increased in OA patients with low-grade inflammation, Ramasamy et al. conducted a pilot study evaluating the interplay between microbiome, knee OA (KOA), and VitD levels, with the ultimate object of establishing whether VitD deficiency is associated with microbiome dysfunction. In the cohort, specific bacteria were identified in VDD subjects compared to normal controls, and KOA_VDD patients showed specific bacteria species that were not identified in KOA without VDD, suggesting a possible influence of VitD on gut microbiome, depending on the KOA status, and a microbiome influence of the KOA condition in VDD subjects. However, in the KOA_VDD group, the core bacteria from KOA or VDD were not identified, suggesting a possible interplay of both conditions in microbiota dysbiosis [116]. Keeping in line with the potential role of VitD supplementation on microbiota composition in patients with different chronic conditions, Kanhere M et al. demonstrated a different gut microbiota composition in patients with VDD and cystic fibrosis compared to patients with normal VD levels; more specifically, potentially pathogenic species were identified in the microbiome of VDD subjects. Moreover, favorable changes in bacteria species were seen in the VDD group after supplementation compared to the placebo group [117]. The effect of VitD supplementation on gut microbiota in healthy individuals has also been studied. In their RCT, Singh et al. showed a significant increment of health-promoting gut probiotics after VitD supplementation [118]. This finding supports the potentially beneficial role of VitD supplementation in microbiota health. Considering how much VitD is related to several chronic inflammatory diseases and how negatively low-grade chronic inflammation impacts general health, adequate VitD storage should be a priority.

Table 1. Clinical studies reporting the effects of Vitamin D supplementation on hallmarks of aging.

Hallmark	Study	Cohort	Aim	Conclusion	Refs.
Genomic instability	Placebo-controlled	92 patients with colorectal adenoma	Evaluate the effect of VitD supplement with 800 UI/day for 6 months	Significant increase in Bax expression (apoptosis promoter); no increase of Bcl-2 expression (apoptosis inhibitor) in VitD group	[61]
	Randomized Clinical Trial	92 subjects with VitD insufficiency, with and without T2DM	Evaluate the effect of VitD supplement with 2000 UI/day for 3 months	Decreased percentage of DNA damage and oxidative parameters when compared to the control group	[62]
Telomere Attrition	Twins UK Cohort ObservationalRegistry	2160 women (aged 18–79) from population-based cohort of twins	Evaluate the relationship between VitD concentrations and the rate of telomere attrition in leukocytes	Positive correlation between serum 25(OH)D levels and telomere length	[64]
	Nurses' Health Study Observational Registry	1424 women	Evaluate the association between both 25(OH)D and 1,25(OH)2D and leukocyte telomere length	Positive correlation between serum 25(OH)D concentrations and leukocyte telomere length	[65]
	Cross sectional analysis from the Health Professional Follow-Up study	2843 men (from studies of telomeres and cancer)	Evaluate the association between 25(OH)D, 1,25-dihydroxyvitamin D (1,25(OH)2D) and leukocyte telomere length	No association between VitD levels and leukocyte telomere length	[66]
	Cross sectional analysis from Northern Finland Birth Cohort 1966	5096 younger adults	Evaluate associations between 25(OH)D and body mass index (BMI) with leukocyte telomere length and whether associations are independent of concentrations of C-reactive protein	No association between VitD levels and leukocyte telomere length	[67]

Table 1. *Cont.*

Hallmark	Study	Cohort	Aim	Conclusion	Refs.
Telomere Attrition	Cohort from the Newcastle 85+ Study	775 older adults (>85 ys), community dwelling, and institutionalized older patients	Evaluate the association between serum VitD concentration and telomere length in blood cells at baseline, 18 and 36 months	Significant positive association between VitD and telomere length at baseline, but inconsistent relationships at subsequent time points	[68]
	Case–control study	African American Gullah women with systemic lupus erythematosus (n = 59) and healthy controls (n = 59)	Evaluate the relationships between VitD status, telomere length, and anti-telomere antibodies	Significant association between shorter telomeres and lower 25(OH)D levels in patients and healthy control; shorter telomeres at follow-up in VitD-deficient patients	[69]
	Retrospective case–control study	62 stable hemodialysis patients and 60 controls	Evaluate the potential protective role of VitD supplementation on telomere length in peripheral mononuclear cells	Hemodialysis patients treated with VitD had greater telomere length in peripheral mononuclear cells	[70]
	Double-blind, placebo-controlled clinical trial	37 overweight African American subjects (18 placebo and 19 intervention)	Evaluate the effect of VitD3 oral supplementation (60,000 IU/month)	VitD supplementation significantly increased telomerase activity	[74]
Epigenetic alteration	Sub analysis from RCT	51overweight/obese African Americans VitD-deficient subjects	Evaluate different effects of VitD3 supplement (600 IU/day, 2000 IU/day, and 4000 IU/day) for 16 weeks compared to placebo	DNAm age was correlated with chronological age in patients with low VitD levels, while the supplementation was able to decrease DNA metilation and slowed epigenetic aging	[75]
	Southampton Women's Survey: subgroup analysis from prospective cohort	230 children from the Southampton Women's Survey (SWS) mother–offspring study	Evaluate the correlation between VitD status and Retinoid-X receptor-alpha (RXRA) methylation in umbilical cord DNA	Methylation at RXRA locus in the umbilical cord was associated with the offspring bone mass; specific methylation at one locus was associated with maternal VitD deficiency	[78]
	The Maternal VitD Osteoporosis Study (MAVIDOS RCT): subgroup analysis	479 received VitD3 supplementation and 486 were in the placebo group	Evaluate DNAm at a specific site in the RXRA locus (10 CpG)	In the cohort of mothers' receiving VitD, the methylation levels in the RXRA-specific site were significantly lower compared to the placebo group	[79]
	Randomized controlled pilot study	Pregnant women receiving VitD3 400 IU (n = 6, control group) or 3800 IU (n = 7, intervention group)	Identify differentially methylated loci among infants born to mothers supplemented with VitD 400 IU/day (i.e., control) versus 3800 IU/day (i.e., intervention)	Compared to control, intervention was associated with mean 25(OH)D levels reflective of sufficient maternal VitD status by birth, as well as significant gains in maternal leukocyte DNA methylation associated with genes involved in cell migration/motility, development, and growth	[80]

Table 1. *Cont.*

Hallmark	Study	Cohort	Aim	Conclusion	Refs.
Cellular senescence	Case–control study	120 individuals (30 young, 30 older, and 30 older in treatment group 1 (1 sachet of Difensil® IMMUNO/day for 12 weeks) and 20 older in treatment group 2 (2 sachet of Difensil® IMMUNO/day for 6 weeks)	Evaluate the effect of the nutraceutical supplement (comprehensive of Sambucus nigra, zinc, tyndallized Lactobacillus acidophilus (HA122), arabinogalactans, VitD, vitamin E, vitamin C, and group B vitamins) on specific signatures of immunosenescence	Older people treated for 30 days improved IL-6, CRP, and lymphocytes levels independent from the dosage of the supplements used; despite the improvement, they were not able to reach the same conditions of young patients	[95]
	Double Blind RCT	110 older fit or frail subjects	Evaluate the effect of alphacalcidol supplementation in inflammatory profile (Il-6, IL-10, TNF, CD4/CD8 ratio, CD8+, and CD28-)	Cohort receiving alphacalcidiol had a better immune profile across the whole cohort as well as the cohort divided according to frailty status	[96]
Chronic inflammation	VITAL study sub-cohort	25,871 women aged ≥55 and men aged ≥50	Measure the effect of VitD supplementation (2000 UI/day) and/or *n*-3 FAs on systemic inflammatory biomarkers	VitD supplementation with or without *n*-3 FAs decreased hs-CRP by 19% at year 2, and the reduction was attenuated at 4 years Other inflammatory biomarkers (IL-6, IL-10, and TNF-α) were not significantly modified	[114]
Dysbiosis	Sub analysis from a RCT	167 HIV-1 patients receiving VitD 5000 IU/day and 500 mg phenylbutyrate (PBA) for 16 weeks	VitD or PBA may have a modulatory action on microbiota composition, metabolites production, and immune activation	No significant improvement in circulating antimicrobial peptide LL-37 and an actual impact on microbiota	[115]
	Pilot study	24 patients with knee osteoarthritis (KOA), healthy controls, with and without VitD deficiency	Evaluate the interplay between microbiome, knee OA (KOA), and VitD levels for establishing whether VitD deficiency is associated with microbiome dysfunction	Patients with KOA and VitD-deficient showed specific bacteria species, absent in KOA, and not VitD-deficient; The different core bacteria in KOA and VitD-deficient suggested an interplay between conditions	[116]
	RCT	41 patients with cystic fibrosis divided in VitD insufficient ($n = 23$) and VitD sufficient ($n = 18$)	Comparing microbiota composition in VitD insufficient and sufficient patients and evaluate the impact of VitD supplement.	Potentially pathogenic species were identified in the microbiome of VitD-deficient subjects	[117]
	Interventional study	100 healthy women	Evaluate gut microbiota composition before and after VitD supplementation	Significant increment of health-promoting gut probiotics after VitD supplementation	[118]

5. Discussion and Future Perspectives

Aging is a complex and multifactorial process, marked by the progressive decline in organ functions, leading to frailty and increased susceptibility to age-related diseases. Although years of research have supported the importance of each hallmark in the process of aging, a significant amount of recent evidence indicates that the hallmarks are not acting in isolation but rather have complex, interconnected interactions. Hallmarks of aging are a set of interconnected biological mechanisms that, individually or synergistically, contribute to molecular and cellular damage, potentially initiating and accelerating the onset of age-related diseases. The interconnection among hallmarks of aging highlights the need to consider aging holistically and investigate biological and phenotypical changes integrally, ultimately disentangling the potential agonistic, antagonistic, and modulator factors leading to organ disease. Currently, we are not certain what impact hallmark interactions have on normal or pathological aging or if there are possible hierarchies among the hallmarks. In addition, whether there is a threshold beyond which such interactions could overwhelm compensatory mechanisms needs to be investigated [15]. Given the epidemiological scenarios and the available scientific and technological achievements, it seems pivotal to move and integrate the routes of research from the biological to the clinical layers in the continuum of dynamism from integrity to damage, from vitality to frailty, with the ultimate goal of expanding a valuable lifespan. In this perspective, it appears a priority to explore the efficacy of available multi-target interventions with potential for modulating the hallmarks of aging, then verifying the impact on the overall body's homeostasis and functioning. Indeed, the hallmarks of aging could represent a milestone to be addressed and understood in healthy aging studies and interventions, hopefully driving the transition from preclinical contexts to clinical settings.

In light of promoting active longevity, strong evidence supports the deleterious effects of hypoVitD on musculoskeletal health, particularly in causing falls, fragility fractures, and mobility issues. Beyond its musculoskeletal effects, VitD may exert several potential extra-skeletal actions, modulating several biomolecular mechanisms and pathways that have been identified as VDR-related in a wide range of cellular types and at multiple cellular layers. Among the biomolecular mechanisms, the hallmarks of aging are pivotal to disentangle the modulatory or causal effects of VitD on the onset and progression of aging and age-related diseases. VitD appears to regulate and influence several hallmarks of aging through genomic and non-genomic pathways, influencing their expression directly or indirectly by activating epigenetic changes or just influencing the cellular environmental milieu. Based on the available clinical studies, we can conclude that ViD may contribute to genomic stability, but it does not seem to have an impact on telomere length, although some findings support a protective role of VitD on telomerase activity in obese individuals. Encouraging evidence supports the epigenetic effects of VitD, which may be able to control and attenuate DNA damage in adults, infants, and their mothers. Then, some evidence supports the protective role of VitD in modulating senescence and related-SASP, but proves concerning the capacity of VitD to lower chronic inflammation in the long term are still waiting to come, as well as those showing better intercellular communication. Calcium is also recognized as a potential regulator of senescence in vivo and plays a key role in several processes in which senescence is involved, such as wound healing, cell migration, and invasion, as well as insulin sensitivity [119], suggesting that calcium and senescence may be functionally linked in many contexts [120]. To date, there is no evidence for an impact of VitD on attenuating SC exhaustion, but it looks promising that VitD modulates the vitality and efficacy of SC infusion. However, caution is needed in interpreting these findings, considering the novelty and the debate about this approach. Notably, clinical studies support the potential of VitD supplementation in hampering dysbiosys, prompting the recovery of gut microbiota health, and providing multiple health benefits.

Despite the interest in VitD supplementation as a strategy supporting human longevity and some evidence about its potential in modulating hallmarks of aging, we are still far from the point of translation from bench to bed. With some exceptions, the majority of human studies investigated the impact of VitD supplementation on the single hallmark of aging by using disease models or peculiar subgroups of the population, focusing mainly on inflammatory-related changes without exploring the correlations with other hallmarks or functional or biological markers of aging. Therefore, current knowledge does not allow yet to define a potential optimal VitD dosage for modulating hallmarks of aging or the chronological or functional age at which VitD supplementation might be useful in modulating aging biologically related factors. However, aging is a progressive process that starts soon after achieving the structural and functional body's maturation. It mainly consists of adaptative changes that lead over time to the accumulation of deficits and insults. Then, we could speculate that early interventions to avoid VitD insufficiency might be protective in the long run, but longitudinal studies based on homogenous cohorts and targeted therapy are warranted.

Indeed, there are few intervention studies conducted in humans that confirm potential pathways already explored in preclinical models. Human studies carried out so far are mainly transverse, and those few longitudinal studies were too small, limited by a short follow-up time, or focused on a single biomolecular mechanism, unable to obtain clarifying results or provide a wide perspective on the overall cellular and pathophysiological changes. Moreover, many literature discrepancies exist and may be justified by the use of different VitD supplements, different dosages, heterogeneous VitD baseline levels, often even unknown, different measurement techniques for hallmarks, and, ultimately, different target populations and aims.

From a clinical standpoint, it is a relevant approach to preventing VitD deficiency at any age since it is associated with osteomalacia and ensuring a sufficient level of VitD in the general population. It is strongly recommended that seniors with a fall history or those living in long-term care institutions maintain or regain optimal VitD levels by means of supplementation (at least 800–1000 IU daily). However, optimal VitD levels are recommended in adults or older adults at high risk for fragility fractures. Supplementation with VitD is a first-line intervention in the secondary prevention of falls and fragility fractures, and, in these cases, it is advisable to measure the patient's serum VitD concentrations after two-to-three months of ViD supplementation. Indeed, there is no direct evidence supporting a clear advantage in performing a basal assessment of VitD at the levels reached with supplementation. On the contrary, the failure or inadequate supplementation with vitamin D in patients on antiresorptive or anabolic treatment for osteoporosis is thought to further amplify the problem and expose patients to a high risk of re-fracture and mortality [121].

In summary, a growing body of research has explored the relationship between VitD and some hallmarks of aging, with varying results depending on study subgroups and design. Despite encouraging routes for potential interventions based on VitD, further investigations are required to elucidate the effects from an integrated perspective and set the pillars for the potential clinical implications of these findings.

Funding: This research received no external funding.

Conflicts of Interest: The authors declare no conflict of interest.

References

1. Lopez-Otin, C.; Blasco, M.A.; Partridge, L.; Serrano, M.; Kroemer, G. The hallmarks of aging. *Cell* **2013**, *153*, 1194–1217. [CrossRef]
2. Lopez-Otin, C.; Blasco, M.A.; Partridge, L.; Serrano, M.; Kroemer, G. Hallmarks of aging: An expanding universe. *Cell* **202** *186*, 243–278. [CrossRef]
3. Tosato, M.; Zamboni, V.; Ferrini, A.; Cesari, M. The aging process and potential interventions to extend life expectancy. *Cli Interv. Aging* **2007**, *2*, 401–412. [PubMed]
4. Polidori, M.C. Aging hallmarks, biomarkers, and clocks for personalized medicine: (Re)positioning the limelight. *Free Radic. Bi Med.* **2024**, *215*, 48–55. [CrossRef]
5. Kennedy, B.K.; Berger, S.L.; Brunet, A.; Campisi, J.; Cuervo, A.M.; Epel, E.S.; Franceschi, C.; Lithgow, G.J.; Morimoto, R.I Pessin, J.E.; et al. Geroscience: Linking aging to chronic disease. *Cell* **2014**, *159*, 709–713. [CrossRef]

6. Bouillon, R.; Marcocci, C.; Carmeliet, G.; Bikle, D.; White, J.H.; Dawson-Hughes, B.; Lips, P.; Munns, C.F.; Lazaretti-Castro, M.; Giustina, A.; et al. Skeletal and Extraskeletal Actions of Vitamin D: Current Evidence and Outstanding Questions. *Endocr. Rev.* **2019**, *40*, 1109–1151. [CrossRef] [PubMed]
7. Luiz, M.M.; Maximo, R.; Oliveira, D.C.; Ramirez, P.C.; de Souza, A.F.; Delinocente, M.L.B.; Steptoe, A.; de Oliveira, C.; Alexandre, T. Association of Serum 25-Hydroxyvitamin D Deficiency with Risk of Incidence of Disability in Basic Activities of Daily Living in Adults >50 Years of Age. *J. Nutr.* **2020**, *150*, 2977–2984. [CrossRef] [PubMed]
8. Luiz, M.M.; Maximo, R.O.; de Oliveira, D.C.; Ramirez, P.C.; de Souza, A.F.; Delinocente, M.L.B.; Steptoe, A.; de Oliveira, C.; Alexandre, T.D.S. Sex Differences in Vitamin D Status as a Risk Factor for Incidence of Disability in Instrumental Activities of Daily Living: Evidence from the ELSA Cohort Study. *Nutrients* **2022**, *14*, 2012. [CrossRef]
9. Zamboni, M.; Zoico, E.; Tosoni, P.; Zivelonghi, A.; Bortolani, A.; Maggi, S.; Di Francesco, V.; Bosello, O. Relation between vitamin D, physical performance, and disability in elderly persons. *J. Gerontol. A Biol. Sci. Med. Sci.* **2002**, *57*, M7–M11. [CrossRef]
10. Girgis, C.M.; Mokbel, N.; Digirolamo, D.J. Therapies for musculoskeletal disease: Can we treat two birds with one stone? *Curr. Osteoporos. Rep.* **2014**, *12*, 142–153. [CrossRef]
11. National Institute of Health and Care Excellence. Vitamin D: Supplement Use in Specific Population Groups. Available online: https://www.nice.org.uk/guidance/ph56 (accessed on 30 December 2023).
12. Gallagher, J.C. Vitamin D and aging. *Endocrinol. Metab. Clin. N. Am.* **2013**, *42*, 319–332. [CrossRef]
13. Vetter, V.M.; Sommerer, Y.; Kalies, C.H.; Spira, D.; Bertram, L.; Demuth, I. Vitamin D supplementation is associated with slower epigenetic aging. *Geroscience* **2022**, *44*, 1847–1859. [CrossRef] [PubMed]
14. Pal, E.; Ungvari, Z.; Benyo, Z.; Varbiro, S. Role of Vitamin D Deficiency in the Pathogenesis of Cardiovascular and Cerebrovascular Diseases. *Nutrients* **2023**, *15*, 334. [CrossRef]
15. Tenchov, R.; Sasso, J.M.; Wang, X.; Zhou, Q.A. Aging Hallmarks and Progression and Age-Related Diseases: A Landscape View of Research Advancement. *ACS Chem. Neurosci.* **2024**, *15*, 1–30. [CrossRef]
16. Zampino, M.; AlGhatrif, M.; Kuo, P.L.; Simonsick, E.M.; Ferrucci, L. Longitudinal Changes in Resting Metabolic Rates with Aging Are Accelerated by Diseases. *Nutrients* **2020**, *12*, 3061. [CrossRef]
17. Blokzijl, F.; de Ligt, J.; Jager, M.; Sasselli, V.; Roerink, S.; Sasaki, N.; Huch, M.; Boymans, S.; Kuijk, E.; Prins, P.; et al. Tissue-specific mutation accumulation in human adult stem cells during life. *Nature* **2016**, *538*, 260–264. [CrossRef]
18. Usman, M.; Woloshynowych, M.; Britto, J.C.; Bilkevic, I.; Glassar, B.; Chapman, S.; Ford-Adams, M.E.; Desai, A.; Bain, M.; Tewfik, I.; et al. Obesity, oxidative DNA damage and vitamin D as predictors of genomic instability in children and adolescents. *Int. J. Obes.* **2021**, *45*, 2095–2107. [CrossRef] [PubMed]
19. Giardini, M.A.; Segatto, M.; da Silva, M.S.; Nunes, V.S.; Cano, M.I. Telomere and telomerase biology. *Prog. Mol. Biol. Transl. Sci.* **2014**, *125*, 1–40. [CrossRef] [PubMed]
20. Raynaud, C.M.; Sabatier, L.; Philipot, O.; Olaussen, K.A.; Soria, J.C. Telomere length, telomeric proteins and genomic instability during the multistep carcinogenic process. *Crit. Rev. Oncol. Hematol.* **2008**, *66*, 99–117. [CrossRef]
21. Gruber, H.J.; Semeraro, M.D.; Renner, W.; Herrmann, M. Telomeres and Age-Related Diseases. *Biomedicines* **2021**, *9*, 1335. [CrossRef] [PubMed]
22. Zarei, M.; Zarezadeh, M.; Hamedi Kalajahi, F.; Javanbakht, M.H. The Relationship Between Vitamin D and Telomere/Telomerase: A Comprehensive Review. *J. Frailty Aging* **2021**, *10*, 2–9. [CrossRef] [PubMed]
23. Munoz-Najar, U.; Sedivy, J.M. Epigenetic control of aging. *Antioxid. Redox Signal* **2011**, *14*, 241–259. [CrossRef] [PubMed]
24. Wang, K.; Liu, H.; Hu, Q.; Wang, L.; Liu, J.; Zheng, Z.; Zhang, W.; Ren, J.; Zhu, F.; Liu, G.H. Epigenetic regulation of aging: Implications for interventions of aging and diseases. *Signal Transduct. Target. Ther.* **2022**, *7*, 374. [CrossRef] [PubMed]
25. Vetter, V.M.; Spira, D.; Banszerus, V.L.; Demuth, I. Epigenetic Clock and Leukocyte Telomere Length Are Associated with Vitamin D Status but not with Functional Assessments and Frailty in the Berlin Aging Study II. *J. Gerontol. A Biol. Sci. Med. Sci.* **2020**, *75*, 2056–2063. [CrossRef] [PubMed]
26. Mark, K.A.; Dumas, K.J.; Bhaumik, D.; Schilling, B.; Davis, S.; Oron, T.R.; Sorensen, D.J.; Lucanic, M.; Brem, R.B.; Melov, S.; et al. Vitamin D Promotes Protein Homeostasis and Longevity via the Stress Response Pathway Genes skn-1, ire-1, and xbp-1. *Cell Rep.* **2016**, *17*, 1227–1237. [CrossRef]
27. Bhutia, S.K. Vitamin D in autophagy signaling for health and diseases: Insights on potential mechanisms and future perspectives. *J. Nutr. Biochem.* **2022**, *99*, 108841. [CrossRef] [PubMed]
28. Zhang, W.B.; Aleksic, S.; Gao, T.; Weiss, E.F.; Demetriou, E.; Verghese, J.; Holtzer, R.; Barzilai, N.; Milman, S. Insulin-like Growth Factor-1 and IGF Binding Proteins Predict All-Cause Mortality and Morbidity in Older Adults. *Cells* **2020**, *9*, 1368. [CrossRef] [PubMed]
29. Chen, Y.; Hamidu, S.; Yang, X.; Yan, Y.; Wang, Q.; Li, L.; Oduro, P.K.; Li, Y. Dietary Supplements and Natural Products: An Update on Their Clinical Effectiveness and Molecular Mechanisms of Action During Accelerated Biological Aging. *Front. Genet.* **2022**, *13*, 880421. [CrossRef] [PubMed]
30. Healthy Ageing and Functional Ability. Available online: https://www.who.int/news-room/questions-and-answers/item/healthy-ageing-and-functional-ability (accessed on 30 December 2023).
31. Gerencser, A.A.; Doczi, J.; Torocsik, B.; Bossy-Wetzel, E.; Adam-Vizi, V. Mitochondrial swelling measurement in situ by optimized spatial filtering: Astrocyte-neuron differences. *Biophys. J.* **2008**, *95*, 2583–2598. [CrossRef] [PubMed]

32. Seo, A.Y.; Joseph, A.M.; Dutta, D.; Hwang, J.C.; Aris, J.P.; Leeuwenburgh, C. New insights into the role of mitochondria in aging: Mitochondrial dynamics and more. *J. Cell Sci.* **2010**, *123*, 2533–2542. [CrossRef]
33. Srivastava, S. The Mitochondrial Basis of Aging and Age-Related Disorders. *Genes* **2017**, *8*, 398. [CrossRef] [PubMed]
34. Chee, C.; Shannon, C.E.; Burns, A.; Selby, A.L.; Wilkinson, D.; Smith, K.; Greenhaff, P.L.; Stephens, F.B. Increasing skeletal muscle carnitine content in older individuals increases whole-body fat oxidation during moderate-intensity exercise. *Aging Cell* **2021**, *20*, e13303. [CrossRef] [PubMed]
35. Coradduzza, D.; Congiargiu, A.; Chen, Z.; Cruciani, S.; Zinellu, A.; Carru, C.; Medici, S. Humanin and Its Pathophysiological Roles in Aging: A Systematic Review. *Biology* **2023**, *12*, 558. [CrossRef] [PubMed]
36. Kim, K.H.; Son, J.M.; Benayoun, B.A.; Lee, C. The Mitochondrial-Encoded Peptide MOTS-c Translocates to the Nucleus to Regulate Nuclear Gene Expression in Response to Metabolic Stress. *Cell Metab.* **2018**, *28*, 516–524.e517. [CrossRef] [PubMed]
37. Matta Reddy, A.; Iqbal, M.; Chopra, H.; Urmi, S.; Junapudi, S.; Bibi, S.; Kumar Gupta, S.; Nirmala Pangi, V.; Singh, I.; Abdel-Daim, M.M. Pivotal role of vitamin D in mitochondrial health, cardiac function, and human reproduction. *EXCLI J.* **2022**, *21*, 967–990. [CrossRef]
38. Hernandez-Segura, A.; Nehme, J.; Demaria, M. Hallmarks of Cellular Senescence. *Trends Cell Biol.* **2018**, *28*, 436–453. [CrossRef] [PubMed]
39. Diwan, B.; Sharma, R. Nutritional components as mitigators of cellular senescence in organismal aging: A comprehensive review. *Food Sci. Biotechnol.* **2022**, *31*, 1089–1109. [CrossRef]
40. Chaib, S.; Tchkonia, T.; Kirkland, J.L. Cellular senescence and senolytics: The path to the clinic. *Nat. Med.* **2022**, *28*, 1556–1568. [CrossRef] [PubMed]
41. Teale, M.A.; Schneider, S.; Eibl, D.; van den Bos, C.; Neubauer, P.; Eibl, R. Mesenchymal and induced pluripotent stem cell-based therapeutics: A comparison. *Appl. Microbiol. Biotechnol.* **2023**, *107*, 4429–4445. [CrossRef]
42. Brunet, A.; Goodell, M.A.; Rando, T.A. Ageing and rejuvenation of tissue stem cells and their niches. *Nat. Rev. Mol. Cell Biol.* **2023**, *24*, 45–62. [CrossRef]
43. Garay, R.P. Recent clinical trials with stem cells to slow or reverse normal aging processes. *Front. Aging* **2023**, *4*, 1148926. [CrossRef]
44. Fantini, C.; Corinaldesi, C.; Lenzi, A.; Migliaccio, S.; Crescioli, C. Vitamin D as a Shield against Aging. *Int. J. Mol. Sci.* **2023**, *24*, 4546. [CrossRef]
45. Baechle, J.J.; Chen, N.; Makhijani, P.; Winer, S.; Furman, D.; Winer, D.A. Chronic inflammation and the hallmarks of aging. *Mol. Metab.* **2023**, *74*, 101755. [CrossRef]
46. Korf, H.; Wenes, M.; Stijlemans, B.; Takiishi, T.; Robert, S.; Miani, M.; Eizirik, D.L.; Gysemans, C.; Mathieu, C. 1,25-Dihydroxyvitamin D3 curtails the inflammatory and T cell stimulatory capacity of macrophages through an IL-10-dependent mechanism. *Immunobiology* **2012**, *217*, 1292–1300. [CrossRef]
47. Ghosh, T.S.; Shanahan, F.; O'Toole, P.W. The gut microbiome as a modulator of healthy ageing. *Nat. Rev. Gastroenterol. Hepatol.* **2022**, *19*, 565–584. [CrossRef]
48. Depommier, C.; Everard, A.; Druart, C.; Plovier, H.; Van Hul, M.; Vieira-Silva, S.; Falony, G.; Raes, J.; Maiter, D.; Delzenne, N.M.; et al. Supplementation with Akkermansia muciniphila in overweight and obese human volunteers: A proof-of-concept exploratory study. *Nat. Med.* **2019**, *25*, 1096–1103. [CrossRef] [PubMed]
49. Ratiner, K.; Ciocan, D.; Abdeen, S.K.; Elinav, E. Utilization of the microbiome in personalized medicine. *Nat. Rev. Microbiol.* **2023**. [CrossRef] [PubMed]
50. Warren, T.; McAllister, R.; Morgan, A.; Rai, T.S.; McGilligan, V.; Ennis, M.; Page, C.; Kelly, C.; Peace, A.; Corfe, B.M.; et al. The Interdependency and Co-Regulation of the Vitamin D and Cholesterol Metabolism. *Cells* **2021**, *10*, 2007. [CrossRef]
51. Mizwicki, M.T.; Keidel, D.; Bula, C.M.; Bishop, J.E.; Zanello, L.P.; Wurtz, J.M.; Moras, D.; Norman, A.W. Identification of an alternative ligand-binding pocket in the nuclear vitamin D receptor and its functional importance in 1alpha,25(OH)2-vitamin D3 signaling. *Proc. Natl. Acad. Sci. USA* **2004**, *101*, 12876–12881. [CrossRef] [PubMed]
52. Chishimba, L.; Thickett, D.R.; Stockley, R.A.; Wood, A.M. The vitamin D axis in the lung: A key role for vitamin D-binding protein. *Thorax* **2010**, *65*, 456–462. [CrossRef] [PubMed]
53. Dancer, R.C.; Parekh, D.; Lax, S.; D'Souza, V.; Zheng, S.; Bassford, C.R.; Park, D.; Bartis, D.G.; Mahida, R.; Turner, A.M.; et al. Vitamin D deficiency contributes directly to the acute respiratory distress syndrome (ARDS). *Thorax* **2015**, *70*, 617–624. [CrossRef]
54. Krasniqi, E.; Boshnjaku, A.; Wagner, K.H.; Wessner, B. Association between Polymorphisms in Vitamin D Pathway-Related Genes, Vitamin D Status, Muscle Mass and Function: A Systematic Review. *Nutrients* **2021**, *13*, 3109. [CrossRef]
55. Bollen, S.E.; Bass, J.J.; Fujita, S.; Wilkinson, D.; Hewison, M.; Atherton, P.J. The Vitamin D/Vitamin D receptor (VDR) axis in muscle atrophy and sarcopenia. *Cell Signal* **2022**, *96*, 110355. [CrossRef]
56. Pusceddu, I.; Farrell, C.J.; Di Pierro, A.M.; Jani, E.; Herrmann, W.; Herrmann, M. The role of telomeres and vitamin D in cellular aging and age-related diseases. *Clin. Chem. Lab. Med.* **2015**, *53*, 1661–1678. [CrossRef]
57. Janubova, M.; Zitnanova, I. The effects of vitamin D on different types of cells. *Steroids* **2024**, *202*, 109350. [CrossRef]
58. Lestari, M.I.; Murti, K.; Liberty, I.A.; Hafy, Z.; Linardi, V.; Khoirudin, M.; Umar, T.P. Association of vitamin D with deoxyribonucleic acid (dna) damage: A systematic review of animal and human studies. *Acta Biochim. Pol.* **2023**, *70*, 379–387. [CrossRef]
59. Nair-Shalliker, V.; Armstrong, B.K.; Fenech, M. Does vitamin D protect against DNA damage? *Mutat. Res.* **2012**, *733*, 50–57. [CrossRef]

60. Graziano, S.; Johnston, R.; Deng, O.; Zhang, J.; Gonzalo, S. Vitamin D/vitamin D receptor axis regulates DNA repair during oncogene-induced senescence. *Oncogene* **2016**, *35*, 5362–5376. [CrossRef] [PubMed]
61. Fedirko, V.; Bostick, R.M.; Long, Q.; Flanders, W.D.; McCullough, M.L.; Sidelnikov, E.; Daniel, C.R.; Rutherford, R.E.; Shaukat, A. Effects of supplemental vitamin D and calcium on oxidative DNA damage marker in normal colorectal mucosa: A randomized clinical trial. *Cancer Epidemiol. Biomarkers Prev.* **2010**, *19*, 280–291. [CrossRef]
62. Wenclewska, S.; Szymczak-Pajor, I.; Drzewoski, J.; Bunk, M.; Sliwinska, A. Vitamin D Supplementation Reduces Both Oxidative DNA Damage and Insulin Resistance in the Elderly with Metabolic Disorders. *Int. J. Mol. Sci.* **2019**, *20*, 2891. [CrossRef]
63. Fagundes, G.E.; Macan, T.P.; Rohr, P.; Damiani, A.P.; Da Rocha, F.R.; Pereira, M.; Longaretti, L.M.; Vilela, T.C.; Ceretta, L.B.; Mendes, C.; et al. Vitamin D3 as adjuvant in the treatment of type 2 diabetes mellitus: Modulation of genomic and biochemical instability. *Mutagenesis* **2019**, *34*, 135–145. [CrossRef] [PubMed]
64. Richards, J.B.; Valdes, A.M.; Gardner, J.P.; Paximadas, D.; Kimura, M.; Nessa, A.; Lu, X.; Surdulescu, G.L.; Swaminathan, R.; Spector, T.D.; et al. Higher serum vitamin D concentrations are associated with longer leukocyte telomere length in women. *Am. J. Clin. Nutr.* **2007**, *86*, 1420–1425. [CrossRef] [PubMed]
65. Liu, J.J.; Prescott, J.; Giovannucci, E.; Hankinson, S.E.; Rosner, B.; Han, J.; De Vivo, I. Plasma vitamin D biomarkers and leukocyte telomere length. *Am. J. Epidemiol.* **2013**, *177*, 1411–1417. [CrossRef] [PubMed]
66. Julin, B.; Shui, I.M.; Prescott, J.; Giovannucci, E.L.; De Vivo, I. Plasma vitamin D biomarkers and leukocyte telomere length in men. *Eur. J. Nutr.* **2017**, *56*, 501–508. [CrossRef] [PubMed]
67. Williams, D.M.; Palaniswamy, S.; Sebert, S.; Buxton, J.L.; Blakemore, A.I.; Hypponen, E.; Jarvelin, M.R. 25-Hydroxyvitamin D Concentration and Leukocyte Telomere Length in Young Adults: Findings From the Northern Finland Birth Cohort 1966. *Am. J. Epidemiol.* **2016**, *183*, 191–198. [CrossRef] [PubMed]
68. Hakeem, S.; Mendonca, N.; Aspray, T.; Kingston, A.; Martin-Ruiz, C.; Robinson, L.; Hill, T.R. The Association between 25-Hydroxyvitamin D Concentration and Telomere Length in the Very-Old: The Newcastle 85+ Study. *Nutrients* **2021**, *13*, 4341. [CrossRef]
69. Hoffecker, B.M.; Raffield, L.M.; Kamen, D.L.; Nowling, T.K. Systemic lupus erythematosus and vitamin D deficiency are associated with shorter telomere length among African Americans: A case-control study. *PLoS ONE* **2013**, *8*, e63725. [CrossRef]
70. Borras, M.; Panizo, S.; Sarro, F.; Valdivielso, J.M.; Fernandez, E. Assessment of the potential role of active vitamin D treatment in telomere length: A case-control study in hemodialysis patients. *Clin. Ther.* **2012**, *34*, 849–856. [CrossRef] [PubMed]
71. Chang, K.V.; Chen, Y.C.; Wu, W.T.; Shen, H.J.; Huang, K.C.; Chu, H.P.; Han, D.S. Expression of Telomeric Repeat-Containing RNA Decreases in Sarcopenia and Increases after Exercise and Nutrition Intervention. *Nutrients* **2020**, *12*, 3766. [CrossRef] [PubMed]
72. Maicher, A.; Kastner, L.; Luke, B. Telomeres and disease: Enter TERRA. *RNA Biol.* **2012**, *9*, 843–849. [CrossRef]
73. Mazidi, M.; Mikhailidis, D.P.; Banach, M.; Dehghan, A. Impact of serum 25-hydroxyvitamin D 25(OH) on telomere attrition: A Mendelian Randomization study. *Clin. Nutr.* **2020**, *39*, 2730–2733. [CrossRef]
74. Zhu, H.; Guo, D.; Li, K.; Pedersen-White, J.; Stallmann-Jorgensen, I.S.; Huang, Y.; Parikh, S.; Liu, K.; Dong, Y. Increased telomerase activity and vitamin D supplementation in overweight African Americans. *Int. J. Obes.* **2012**, *36*, 805–809. [CrossRef] [PubMed]
75. Chen, L.; Dong, Y.; Bhagatwala, J.; Raed, A.; Huang, Y.; Zhu, H. Effects of Vitamin D3 Supplementation on Epigenetic Aging in Overweight and Obese African Americans With Suboptimal Vitamin D Status: A Randomized Clinical Trial. *J. Gerontol. A Biol. Sci. Med. Sci.* **2019**, *74*, 91–98. [CrossRef] [PubMed]
76. Zhu, H.; Bhagatwala, J.; Huang, Y.; Pollock, N.K.; Parikh, S.; Raed, A.; Gutin, B.; Harshfield, G.A.; Dong, Y. Race/Ethnicity-Specific Association of Vitamin D and Global DNA Methylation: Cross-Sectional and Interventional Findings. *PLoS ONE* **2016**, *11*, e0152849. [CrossRef] [PubMed]
77. Zhu, H.; Wang, X.; Shi, H.; Su, S.; Harshfield, G.A.; Gutin, B.; Snieder, H.; Dong, Y. A genome-wide methylation study of severe vitamin D deficiency in African American adolescents. *J. Pediatr.* **2013**, *162*, 1004–1009.e1. [CrossRef] [PubMed]
78. Harvey, N.C.; Sheppard, A.; Godfrey, K.M.; McLean, C.; Garratt, E.; Ntani, G.; Davies, L.; Murray, R.; Inskip, H.M.; Gluckman, P.D.; et al. Childhood bone mineral content is associated with methylation status of the RXRA promoter at birth. *J. Bone Miner. Res.* **2014**, *29*, 600–607. [CrossRef] [PubMed]
79. Curtis, E.M.; Krstic, N.; Cook, E.; D'Angelo, S.; Crozier, S.R.; Moon, R.J.; Murray, R.; Garratt, E.; Costello, P.; Cleal, J.; et al. Gestational Vitamin D Supplementation Leads to Reduced Perinatal RXRA DNA Methylation: Results From the MAVIDOS Trial. *J. Bone Miner. Res.* **2019**, *34*, 231–240. [CrossRef] [PubMed]
80. Anderson, C.M.; Gillespie, S.L.; Thiele, D.K.; Ralph, J.L.; Ohm, J.E. Effects of Maternal Vitamin D Supplementation on the Maternal and Infant Epigenome. *Breastfeed. Med.* **2018**, *13*, 371–380. [CrossRef]
81. Bollen, S.E.; Atherton, P.J. Myogenic, genomic and non-genomic influences of the vitamin D axis in skeletal muscle. *Cell Biochem. Funct.* **2021**, *39*, 48–59. [CrossRef]
82. Reis, N.G.; Assis, A.P.; Lautherbach, N.; Goncalves, D.A.; Silveira, W.A.; Morgan, H.J.N.; Valentim, R.R.; Almeida, L.F.; Heck, L.C.; Zanon, N.M.; et al. Maternal vitamin D deficiency affects the morphology and function of glycolytic muscle in adult offspring rats. *J. Cachexia Sarcopenia Muscle* **2022**, *13*, 2175–2187. [CrossRef]
83. Das, A.; Jawla, N.; Meena, V.; Gopinath, S.D.; Arimbasseri, G.A. Lack of vitamin D signalling shifts skeletal muscles towards oxidative metabolism. *J. Cachexia Sarcopenia Muscle* **2023**, *15*, 67–80. [CrossRef] [PubMed]
84. Ricca, C.; Aillon, A.; Bergandi, L.; Alotto, D.; Castagnoli, C.; Silvagno, F. Vitamin D Receptor Is Necessary for Mitochondrial Function and Cell Health. *Int. J. Mol. Sci.* **2018**, *19*, 1672. [CrossRef] [PubMed]

85. Hua, R.; Liu, B.; He, W.; Zhang, H.; Liu, Y.; Xie, Q.; Zhou, L.; Pei, F. Calcitriol reverses age-related hypertension via downregulatin renal AP1/AT(1)R pathway through regulating mitochondrial function. *Clin. Exp. Hypertens.* **2023**, *45*, 2277653. [CrossRef] [PubMed]
86. Mohanad, M.; Mohamed, S.K.; Aboulhoda, B.E.; Ahmed, M.A.E. Neuroprotective effects of vitamin D in an Alzheimer's diseas rat model: Improvement of mitochondrial dysfunction via calcium/calmodulin-dependent protein kinase kinase 2 activation c Sirtuin1 phosphorylation. *Biofactors* **2023**. [CrossRef] [PubMed]
87. Russo, C.; Valle, M.S.; Casabona, A.; Spicuzza, L.; Sambataro, G.; Malaguarnera, L. Vitamin D Impacts on Skeletal Muscl Dysfunction in Patients with COPD Promoting Mitochondrial Health. *Biomedicines* **2022**, *10*, 898. [CrossRef] [PubMed]
88. de Las Heras, N.; Martin Gimenez, V.M.; Ferder, L.; Manucha, W.; Lahera, V. Implications of Oxidative Stress and Potential Rol of Mitochondrial Dysfunction in COVID-19: Therapeutic Effects of Vitamin D. *Antioxidants* **2020**, *9*, 897. [CrossRef]
89. Lee, T.W.; Kao, Y.H.; Chen, Y.J.; Chao, T.F.; Lee, T.I. Therapeutic potential of vitamin D in AGE/RAGE-related cardiovascula diseases. *Cell Mol. Life Sci.* **2019**, *76*, 4103–4115. [CrossRef] [PubMed]
90. Phillips, E.A.; Hendricks, N.; Bucher, M.; Maloyan, A. Vitamin D Supplementation Improves Mitochondrial Function and Reduce Inflammation in Placentae of Obese Women. *Front. Endocrinol.* **2022**, *13*, 893848. [CrossRef]
91. Yu, M.; Wu, H.; Wang, J.; Chen, X.; Pan, J.; Liu, P.; Zhang, J.; Chen, Y.; Zhu, W.; Tang, C.; et al. Vitamin D receptor inhibits EMT vi regulation of the epithelial mitochondrial function in intestinal fibrosis. *J. Biol. Chem.* **2021**, *296*, 100531. [CrossRef]
92. Romani, M.; Berger, M.M.; D'Amelio, P. From the Bench to the Bedside: Branched Amino Acid and Micronutrient Strategies t Improve Mitochondrial Dysfunction Leading to Sarcopenia. *Nutrients* **2022**, *14*, 483. [CrossRef] [PubMed]
93. Martinelli, R.P.; Rayego-Mateos, S.; Alique, M.; Marquez-Exposito, L.; Tejedor-Santamaria, L.; Ortiz, A.; Gonzalez-Parra, E Ruiz-Ortega, M. Vitamin D, Cellular Senescence and Chronic Kidney Diseases: What Is Missing in the Equation? *Nutrients* **202** *15*, 1349. [CrossRef] [PubMed]
94. Yousefzadeh, M.J.; Flores, R.R.; Zhu, Y.; Schmiechen, Z.C.; Brooks, R.W.; Trussoni, C.E.; Cui, Y.; Angelini, L.; Lee, K.A.; McGowa S.J.; et al. An aged immune system drives senescence and ageing of solid organs. *Nature* **2021**, *594*, 100–105. [CrossRef] [PubMed]
95. Maselli Del Giudice, A.; La Mantia, I.; Barbara, F.; Ciccarone, S.; Ragno, M.S.; de Robertis, V.; Cariti, F.; Barbara, M.; D'Ascanio, L Di Stadio, A. Use of Nutraceuticals in Elderly to Fight Inflammation and Immuno-Senescence: A Randomized Case-Contro Study. *Nutrients* **2022**, *14*, 3476. [CrossRef] [PubMed]
96. Rizka, A.; Setiati, S.; Harimurti, K.; Sadikin, M.; Mansur, I.G. Effect of Alfacalcidol on Inflammatory markers and T Cell Subset in Elderly with Frailty Syndrome: A Double Blind Randomized Controlled Trial. *Acta Med. Indones.* **2018**, *50*, 215–221. [PubMed]
97. Martin, N.; Zhu, K.; Czarnecka-Herok, J.; Vernier, M.; Bernard, D. Regulation and role of calcium in cellular senescence. *Ce Calcium* **2023**, *110*, 102701. [CrossRef]
98. Mahdavi, R.; Belgheisi, G.; Haghbin-Nazarpak, M.; Omidi, M.; Khojasteh, A.; Solati-Hashjin, M. Bone tissue engineering gelati hydroxyapatite/graphene oxide scaffolds with the ability to release vitamin D: Fabrication, characterization, and in vitro stud *J. Mater. Sci. Mater. Med.* **2020**, *31*, 97. [CrossRef]
99. Bottai, D.; Fiocco, R.; Gelain, F.; Defilippis, L.; Galli, R.; Gritti, A.; Vescovi, L.A. Neural stem cells in the adult nervous system *J. Hematotherapy Stem Cell Res.* **2003**, *12*, 655–670. [CrossRef] [PubMed]
100. Uthaiah, C.A.; Beeraka, N.M.; Rajalakshmi, R.; Ramya, C.M.; Madhunapantula, S.V. Role of Neural Stem Cells and Vitamin I Receptor (VDR)-Mediated Cellular Signaling in the Mitigation of Neurological Diseases. *Mol. Neurobiol.* **2022**, *59*, 4065–410 [CrossRef]
101. Mouchel, P.L.; Berard, E.; Tavitian, S.; Gadaud, N.; Vergez, F.; Rieu, J.B.; Luquet, I.; Sarry, A.; Huguet, F.; Largeaud, L.; et a Vitamin C and D supplementation in acute myeloid leukemia. *Blood Adv.* **2023**, *7*, 6886–6897. [CrossRef]
102. Paubelle, E.; Zylbersztejn, F.; Maciel, T.T.; Carvalho, C.; Mupo, A.; Cheok, M.; Lieben, L.; Sujobert, P.; Decroocq, J.; Yokoyam A.; et al. Vitamin D Receptor Controls Cell Stemness in Acute Myeloid Leukemia and in Normal Bone Marrow. *Cell Rep.* **202** *30*, 739–754.e734. [CrossRef]
103. Cao, H.; Xu, Y.; de Necochea-Campion, R.; Baylink, D.J.; Payne, K.J.; Tang, X.; Ratanatharathorn, C.; Ji, Y.; Mirshahidi, S.; Che C.S. Application of vitamin D and vitamin D analogs in acute myelogenous leukemia. *Exp. Hematol.* **2017**, *50*, 1–12. [CrossRef]
104. Dantas, J.R.; Araujo, D.B.; Silva, K.R.; Souto, D.L.; de Fatima Carvalho Pereira, M.; Luiz, R.R.; Dos Santos Mantuano, M Claudio-da-Silva, C.; Gabbay, M.A.L.; Dib, S.A.; et al. Adipose tissue-derived stromal/stem cells + cholecalciferol: A pilot stud in recent-onset type 1 diabetes patients. *Arch. Endocrinol. Metab.* **2021**, *65*, 342–351. [CrossRef]
105. Bloemen, V.; Schoenmaker, T.; de Vries, T.J.; Everts, V. Direct cell-cell contact between periodontal ligament fibroblasts an osteoclast precursors synergistically increases the expression of genes related to osteoclastogenesis. *J. Cell Physiol.* **2010**, *222*, 565 573. [CrossRef] [PubMed]
106. Schirrmacher, K.; Bingmann, D. Effects of vitamin D3, 17beta-estradiol, vasoactive intestinal peptide, and glutamate on electri coupling between rat osteoblast-like cells in vitro. *Bone* **1998**, *23*, 521–526. [CrossRef] [PubMed]
107. Sirbe, C.; Rednic, S.; Grama, A.; Pop, T.L. An Update on the Effects of Vitamin D on the Immune System and Autoimmun Diseases. *Int. J. Mol. Sci.* **2022**, *23*, 9784. [CrossRef]
108. Mansournia, M.A.; Ostadmohammadi, V.; Doosti-Irani, A.; Ghayour-Mobarhan, M.; Ferns, G.; Akbari, H.; Ghaderi, A.; Talar H.R.; Asemi, Z. The Effects of Vitamin D Supplementation on Biomarkers of Inflammation and Oxidative Stress in Diabeti Patients: A Systematic Review and Meta-Analysis of Randomized Controlled Trials. *Horm. Metab. Res.* **2018**, *50*, 429–44 [CrossRef] [PubMed]

109. Jamilian, H.; Amirani, E.; Milajerdi, A.; Kolahdooz, F.; Mirzaei, H.; Zaroudi, M.; Ghaderi, A.; Asemi, Z. The effects of vitamin D supplementation on mental health, and biomarkers of inflammation and oxidative stress in patients with psychiatric disorders: A systematic review and meta-analysis of randomized controlled trials. *Prog. Neuropsychopharmacol. Biol. Psychiatry* **2019**, *94*, 109651. [CrossRef] [PubMed]
110. Akbari, M.; Ostadmohammadi, V.; Lankarani, K.B.; Tabrizi, R.; Kolahdooz, F.; Heydari, S.T.; Kavari, S.H.; Mirhosseini, N.; Mafi, A.; Dastorani, M.; et al. The Effects of Vitamin D Supplementation on Biomarkers of Inflammation and Oxidative Stress among Women with Polycystic Ovary Syndrome: A Systematic Review and Meta-Analysis of Randomized Controlled Trials. *Horm. Metab. Res.* **2018**, *50*, 271–279. [CrossRef]
111. Mansouri, L.; Lundwall, K.; Moshfegh, A.; Jacobson, S.H.; Lundahl, J.; Spaak, J. Vitamin D receptor activation reduces inflammatory cytokines and plasma MicroRNAs in moderate chronic kidney disease—A randomized trial. *BMC Nephrol.* **2017**, *18*, 161. [CrossRef]
112. Mousa, A.; Naderpoor, N.; Teede, H.; Scragg, R.; de Courten, B. Vitamin D supplementation for improvement of chronic low-grade inflammation in patients with type 2 diabetes: A systematic review and meta-analysis of randomized controlled trials. *Nutr. Rev.* **2018**, *76*, 380–394. [CrossRef]
113. Autier, P.; Mullie, P.; Macacu, A.; Dragomir, M.; Boniol, M.; Coppens, K.; Pizot, C.; Boniol, M. Effect of vitamin D supplementation on non-skeletal disorders: A systematic review of meta-analyses and randomised trials. *Lancet Diabetes Endocrinol.* **2017**, *5*, 986–1004. [CrossRef]
114. Dong, Y.; Zhu, H.; Chen, L.; Huang, Y.; Christen, W.; Cook, N.R.; Copeland, T.; Mora, S.; Buring, J.E.; Lee, I.M.; et al. Effects of Vitamin D(3) and Marine Omega-3 Fatty Acids Supplementation on Biomarkers of Systemic Inflammation: 4-Year Findings from the VITAL Randomized Trial. *Nutrients* **2022**, *14*, 5307. [CrossRef]
115. Missailidis, C.; Sorensen, N.; Ashenafi, S.; Amogne, W.; Kassa, E.; Bekele, A.; Getachew, M.; Gebreselassie, N.; Aseffa, A.; Aderaye, G.; et al. Vitamin D and Phenylbutyrate Supplementation Does Not Modulate Gut Derived Immune Activation in HIV-1. *Nutrients* **2019**, *11*, 1675. [CrossRef]
116. Ramasamy, B.; Magne, F.; Tripathy, S.K.; Venugopal, G.; Mukherjee, D.; Balamurugan, R. Association of Gut Microbiome and Vitamin D Deficiency in Knee Osteoarthritis Patients: A Pilot Study. *Nutrients* **2021**, *13*, 1272. [CrossRef]
117. Kanhere, M.; He, J.; Chassaing, B.; Ziegler, T.R.; Alvarez, J.A.; Ivie, E.A.; Hao, L.; Hanfelt, J.; Gewirtz, A.T.; Tangpricha, V. Bolus Weekly Vitamin D3 Supplementation Impacts Gut and Airway Microbiota in Adults With Cystic Fibrosis: A Double-Blind, Randomized, Placebo-Controlled Clinical Trial. *J. Clin. Endocrinol. Metab.* **2018**, *103*, 564–574. [CrossRef]
118. Singh, P.; Rawat, A.; Alwakeel, M.; Sharif, E.; Al Khodor, S. The potential role of vitamin D supplementation as a gut microbiota modifier in healthy individuals. *Sci. Rep.* **2020**, *10*, 21641. [CrossRef]
119. Martin, N.; Bernard, D. Calcium signaling and cellular senescence. *Cell Calcium* **2018**, *70*, 16–23. [CrossRef]
120. Terrell, K.; Choi, S.; Choi, S. Calcium's Role and Signaling in Aging Muscle, Cellular Senescence, and Mineral Interactions. *Int. J. Mol. Sci.* **2023**, *24*, 7034. [CrossRef]
121. Bertoldo, F.; Cianferotti, L.; Di Monaco, M.; Falchetti, A.; Fassio, A.; Gatti, D.; Gennari, L.; Giannini, S.; Girasole, G.; Gonnelli, S.; et al. Definition, Assessment, and Management of Vitamin D Inadequacy: Suggestions, Recommendations, and Warnings from the Italian Society for Osteoporosis, Mineral Metabolism and Bone Diseases (SIOMMMS). *Nutrients* **2022**, *14*, 4148. [CrossRef]

Disclaimer/Publisher's Note: The statements, opinions and data contained in all publications are solely those of the individual author(s) and contributor(s) and not of MDPI and/or the editor(s). MDPI and/or the editor(s) disclaim responsibility for any injury to people or property resulting from any ideas, methods, instructions or products referred to in the content.

nutrients

MDPI

Article

Comparative Study of Healthy Older and Younger Adults Shows They Have the Same Skin Concentration of Vitamin D_3 Precursor, 7-Dehydrocholesterol, and Similar Response to UVR

Oktawia Borecka [1,2,3], John J. Dutton [4], Jonathan C. Y. Tang [4,5], William D. Fraser [4,5], Ann R. Webb [1] and Lesley E. Rhodes [2,3,*]

1 Department of Earth and Environmental Sciences, Faculty of Science and Engineering, University of Manchester, Manchester M13 9PL, UK; oktawia.v@gmail.com (O.B.); ann.webb@manchester.ac.uk (A.R.W.)
2 Division of Musculoskeletal and Dermatological Sciences, School of Biological Sciences, Faculty of Biology Medicine and Health, University of Manchester, Manchester M13 9PL, UK
3 Photobiology Unit, Dermatology Research Centre, Salford Royal Hospital, Northern Care Alliance NHS Foundation Trust, Manchester Academic Health Science Centre, Greater Manchester M6 8HD, UK
4 Bioanalytical Facility, Norwich Medical School, University of East Anglia, Norwich Research Park, Norwich NR4 7TJ, UK; jjdutton1@hotmail.co.uk (J.J.D.); jonathan.tang@uea.ac.uk (J.C.Y.T.); w.fraser@uea.ac.uk (W.D.F.)
5 Departments of Clinical Biochemistry and Endocrinology, Norfolk and Norwich University Hospital, Norwich NR4 7UY, UK
* Correspondence: lesley.e.rhodes@manchester.ac.uk

Abstract: Vitamin D_3 synthesis in human skin is initiated by solar ultraviolet radiation (UVR) exposure of precursor 7-dehydrocholesterol (7DHC), but influence of age on the early stage of vitamin D_3 metabolism is uncertain. We performed a prospective standardised study in healthy ambulant adults aged $\geq$65 and $\leq$40 years examining (1) if baseline skin 7DHC concentration differs between younger and older adults and (2) the impact of older age on serum vitamin D_3 response to solar simulated UVR. Eleven younger (18–40 years) and 10 older (65–89 years) adults, phototype I–III, received low-dose UVR (95% UVA, 5% UVB, 1.3 SED) to ~35% of the body surface area. Biopsies were taken for 7DHC assay from unexposed skin, skin immediately and 24 h post-UVR, and blood sampled at baseline, 24 h and 7 d post-UVR for vitamin D_3 assay. Samples were analysed by HPLC-MS/MS. Baseline skin 7DHC (mean $\pm$ SD) was 0.22 $\pm$ 0.07 and 0.25 $\pm$ 0.08 µg/mg in younger versus older adults (no significant difference). Baseline serum vitamin D_3 concentration was 1.5 $\pm$ 1.5 and 1.5 $\pm$ 1.7 nmol/L in younger versus older adults, respectively, and showed a significant increase in both groups post-UVR (no significant differences between age groups). Thus, skin 7DHC concentration was not a limiting factor for vitamin D_3 production in older relative to younger adults. This information assists public health guidance on sun exposure/vitamin D nutrition, with particular relevance to the growing populations of healthy ambulant adults $\geq$65 years.

Keywords: skin; 7-dehydrocholesterol; vitamin D; photobiology; ultraviolet radiation; elderly

1. Introduction

Vitamin D is established to be essential for musculoskeletal health. However, older adults are prone to a low vitamin D status [1]. As there is a growing population of >65-year-olds living in developed countries, including in the UK, where almost a quarter (24.4%) of the population is aged 60 or above [2,3], it is imperative to better understand the functional aspects of ageing skin. The early stage of vitamin D_3 metabolism is one such crucial area for translational research.

Skin contains the precursor for vitamin D_3 (cholecalciferol) synthesis, i.e., 7-dehydrocholesterol (7DHC), with its exposure to solar ultraviolet radiation (UVR) initiating vitamin D_3 biosynthesis. This is usually the major source of vitamin D in humans, with oral intake usually making a smaller contribution [4]. Although both vitamin D_2 and vitamin D_3 can be obtained orally, only vitamin D_3 is synthesised by the skin. The 7DHC is found in higher concentrations in the upper layer of the skin, i.e., the epidermis, than the lower layer, i.e., the dermis, and is present in both keratinocytes and fibroblasts [5,6]. It is rapidly photoconverted to pre-vitamin D_3, with peak production in the ultraviolet B (UVB) waveband at 295 nm [7]. Subsequently, pre-vitamin D_3 is converted more slowly to vitamin D_3 through heat-induced isomerisation. The vitamin D_3 diffuses through the dermal capillary beds and enters the body's general circulation [8]; hence, it is the metabolite that links the skin and blood pathways of vitamin D_3 supply and metabolism. It then undergoes hepatic hydroxylation to 25-hydroxyvitamin D_3 (25(OH)D_3), and following this, renal hydroxylation to the active hormone, i.e., 1,25-dihydroxyvitamin D_3 (1,25(OH)$_2D_3$). Ex vivo studies indicate that less than 15% of 7DHC is converted to pre-vitamin D_3 by a single low UVR exposure [9]. However, higher levels of UVR exposure, and exposure to the ultraviolet A (UVA) component of solar UVR, can, on the other hand, lead to the production of inactive photoisomers, and if prolonged, might degrade newly formed vitamin D_3 [9].

A study published in the 1980s involving skin specimens from surgical patients aged 8–92 years reported a negative correlation ($r = -0.89$) between patient age and epidermal 7DHC concentration, and concluded that a lower vitamin D status in older people could be explained by a lower skin 7DHC concentration [10]. However, the study had some limitations, including the use of skin samples taken opportunistically from different body sites following a variety of surgical procedures; these ranged from leg amputation to reduction surgery and correction of protuberant ears in children [10]. In contrast, a further study using skin from surgical samples and dermatology outpatients (from unspecified body sites) reported no correlation between patient age and skin 7DHC concentration [11]. These earlier studies used liquid chromatography (LC)–UV detection methodology, and no further research addressing skin 7DHC concentration and ageing has to our knowledge been published.

Here, we have performed a standardised comparative study in prospectively recruited groups of younger and older healthy ambulant adults, with the objective firstly to compare the baseline skin content of 7DHC between these age groups, and secondly to examine their 7DHC–vitamin D_3 response to UVR exposure. Multiple sampling was performed of matched skin sites (for 7DHC) and blood (primarily for vitamin D_3) pre and post low, sub-erythemal (sub-sunburn) dose of solar simulated UVR (SSR; emission comprising 5% UVB and 95% UVA) in vivo. This replicates the dose and UVB: UVA balance that can be obtained from ambient solar UVR exposure in everyday life. A modern HPLC–MS/MS assay was employed for analyte measurements, including our recently developed methodology for measurement of 7DHC [12,13].

2. Materials & Methods

2.1. Volunteers

This study was performed in volunteers residing in Greater Manchester, UK, in the Photobiology Unit, Dermatology Research Centre, University of Manchester, based at Salford Royal Hospital, Greater Manchester, UK, in 2019. Inclusion criteria were healthy volunteers, ambulant males or females, of phototype I–III (white Caucasian), and aged 18–40 or 65–89 years. Exclusion criteria were history of skin cancer or photosensitivity, use of a sunbed or sunbathing within the past 3 months, and taking photoactive or bone active therapies, vitamin D > 200 IU (5 μg)/day, or anticoagulants (including aspirin, clopidogrel, warfarin, and propranolol).

Volunteers were recruited through advertisement and the Photobiology Unit database. Four potential volunteers were excluded: three younger adults did not fit the phototype criteria, and one older adult was undergoing immune-suppressant therapy. The North West Greater Manchester Research Ethics Committee gave approval (reference 18/NW/0493) on 25 July 2018, and volunteers completed the study during January–March 2019. The study adhered to Declaration of Helsinki principles, and volunteers gave written, informed consent. The trial is registered on the ISRCTN website, reference ISRCTN72674753.

2.2. Protocol

The study protocol is shown in a flowchart (Figure 1), with further details of the methodology in the following text.

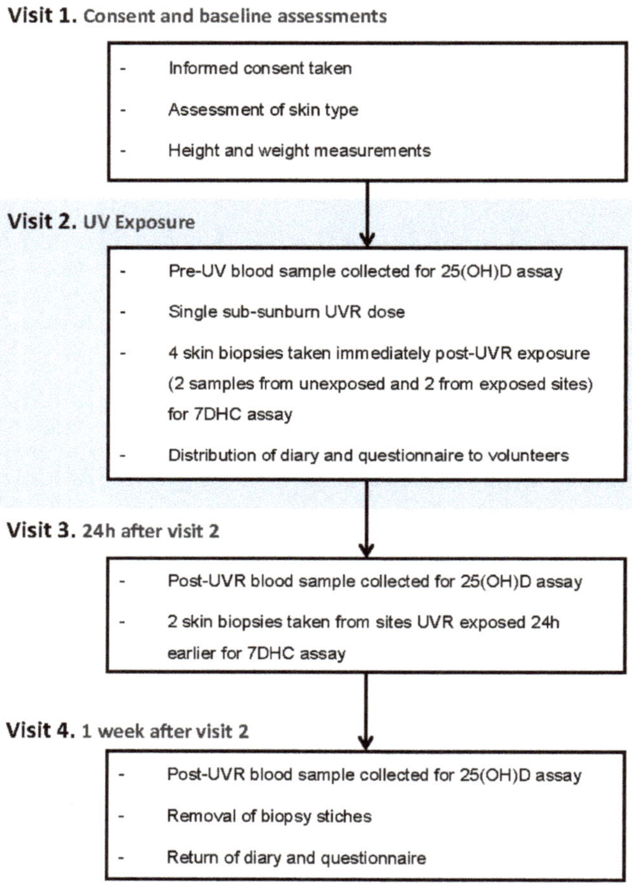

Figure 1. Research protocol flowchart.

2.3. Phototype Assessment

Volunteers were assessed for sun-reactive skin type (phototype) by researchers based at the Photobiology Unit, Salford Royal Hospital, using a standardised series of questions relating to their history of skin sunburn and tanning responses to sunlight exposure, alongside their physical characteristics, including eye, hair, and skin colour (modified Fitzpatrick system [14]).

2.4. Simulated Summer Sunlight Exposures

Volunteers were given a single sub-sunburn SSR dose: 1.3 standard erythemal dose (SED), equivalent to ~15 min UK midday summer exposure. This dose took approximately 6 min 20 s to administer.

A whole-body irradiation cabinet (Philips HB588 Sunstudio, Eindhoven, The Netherlands) was re-fitted with Arimed B fluorescent lamps (Cosmedico GmbH, Stuttgart, Germany), providing UVR emission close to UK ambient midday summer sunlight (290–400 nm; 95% UVA: 320–400 nm, 5% UVB: 290–320 nm). The cabinet emission was characterised with the use of a DTM300 spectroradiometer (Bentham, Reading, UK).

During the UVR exposure, the volunteers lay prone and wore standardised knee-length shorts and a short-sleeved T-shirt, i.e., exposing their hands, forearms, face, and lower legs, equating to ~35% of the body surface area (BSA). A 10 cm × 10 cm area was cut out from one side of their shorts to expose an upper buttock site (UVR biopsy site), and the other buttock was further covered with opaque UVR material (unexposed site). The volunteers wore protective eye goggles.

The study was performed as planned between the months of January and March, in accordance with ambient UVB being insufficient to produce appreciable vitamin D_3 at the study latitude (Manchester, UK; 53.5° N) during that time period. The UV index (UVI, measure of erythemal effective UV radiation) has been monitored in Manchester since 1997, and daily plots of the UVI can be viewed at https://uk-air.defra.gov.uk/data/uv-index-graphs (accessed on 1 February 2020), where a bell-shaped curve indicates a clear day, and interruptions from such are indicative of clouds. When the UVI is less than 2, it is generally accepted as "safe," i.e., minimal risk of sunburn and sun protection not required. Vitamin D synthesis is also deemed negligible in any practical scenario of daily life. The UVI only approaches 2 in the middle of the day at the very end of March, and earlier in the year it is far lower.

2.5. Skin Biopsies and 7DHC Concentration Analysis

Six 5 mm buttock skin punch biopsies were taken from each volunteer, i.e., two from unexposed skin, two immediately post-UVR (within 30 min of the irradiation), and two at 24 h post-UVR. Extraction, chromatographic separation, and measurement of 7DHC in µg was performed using HPLC–MS/MS, as previously reported [12]. In brief, 7DHC was extracted using ethyl acetate:methanol (1:1, v/v) and derivatized with 4-phenyl-1,2,4-triazoline-3,5-dione (PTAD) to enhance the ionization efficiency for electrospray ionization mass spectrometry (ESI–MS). Additionally, solid-supported liquid extraction (SLE) was employed to eliminate larger lipids from the 7DHC and reduce potential matrix effects. The LC–MS/MS assay met the validation criteria set by the International Council for Harmonisation. The calibration curve demonstrated linearity, with an average r2 value of 0.997. The coefficients of variation were 11.1% and 4.32% for inter-assay and intra-assay precision, respectively. The lower limit of quantification was 1.6 µg/g, and the upper limit was 100 µg/g. The average recovery of 7DHC through SLE was 91.4%.

Total cellular protein content of the biopsies was used to normalise the 7DHC assay results. This was performed with a colorimetric Lowry-based protein assay using the DC Protein Assay Kit II (Bio-Rad, Hercules, CA, USA). Samples and the calibration curve were prepared following the manufacturer's protocol [15], and the absorbance was measured using a Multiskan FC plate reader (Thermo Fisher Scientific, Waltham, MA, USA) with an internal shaker at 650 nm. The data obtained from the protein assay were expressed as mg of protein per biopsy, and the 7DHC concentration was expressed as µg of 7DHC per mg of protein.

2.6. Blood Sampling, Analysis of Serum Vitamin D Metabolites, and Routine Biochemistry

Blood samples were collected on 3 occasions, i.e., immediately prior to UVR, and 24 h and 7 days post-UVR, and serum was stored at −80 °C prior to analysis. Analysis of serum vitamin D_3 and vitamin D_2, 25(OH)D_2, and 25(OH)D_3 was performed by LC–MS/MS. Serum 25(OH)D_2 and 25(OH)D_3 were measured using a Micromass Quattro Ultima Pt electrospray ionisation mass spectrometer (Waters Corp., Milford, MA, USA), as described previously [13]. The measurement ranges of the assays were 0.1–200 nmol/L for 25(OH)D_2 and 25(OH)D_3, calibrated using standard reference material SRM972a from the National

Institute of Science and Technology (NIST). The mean coefficient of variation (CV) for intra-assay imprecision across the measuring range of the assays was 4.9% for 25(OH)D_3 and 8.3% for 25(OH)D_3, and the cumulative inter-assay CV was ≤7.4% for 25(OH)D_2 and ≤9.6% for 25(OH)D_3. Our 25(OH)D assays showed <6% accuracy bias against the Centers for Disease Control and Prevention's reference method on the Vitamin D External Quality Assessment Scheme (DEQAS). We met the certification performance standards set by DEQAS throughout the time the analyses were performed. Vitamin D_3 and vitamin D_2 were analysed using the 4-phenyl-1,2,4-triazoline-3,5h-dione (PTAD) derivatised precursor to product mass ion transitions for vitamin D_3 (591.4 > 298.1) and vitamin D_2 (603.5 > 298.1). Serum vitamin D_3 and D_2 assays were calibrated using certified pure standards (IsoSciences, King of Prussia, PA, USA) spiked into vitamin D depleted serum (BBSI solutions, Recklinghausen, Germany) and an isotopic-labelled vitamin D_3-[23,24,25,26,27-13C5] as an internal standard (IsoSciences). The intra- and inter-assay precision coefficients of variation (CV) were 2.2–9.4% across the calibration range of 0–250 nmol/L, with an assay recovery of between 99.7 and 106.3%. The lower limit of quantification (LLoQ) was 1 nmol/L for both vitamin D_3 and vitamin D_2; the lower limit of detection (LLoD) was 0.5 nmol/L. All analyses were undertaken by the Good Clinical Laboratory Practice and DEQAS-certified Bioanalytical Facility at the University of East Anglia. Deficient vitamin D status was defined as 25(OH)D < 25 nmoL/L and sufficient status as 50 to 120 nmol/L [16].

Routine biochemistry, including renal function and parathyroid hormone (PTH), was analysed at Salford Royal Hospital, Greater Manchester, UK. Analysis of PTH was with a 2-site sandwich immunoassay using direct chemiluminometric technology (Siemens Centaur XP Intact PTH assay), and routine biochemistry was analysed via the Siemens ADVIA assay.

2.7. Sample Size and Statistical Analyses

Sample size was constrained by the number of adults willing to undergo studies involving multiple skin biopsies. The sample size was estimated using confidence interval methodology. Based on data on 7DHC concentration in earlier studies [10,11], a sample size of n = 10 in each age group was estimated to have sufficient power to detect an approximate 2-fold difference in baseline skin 7DHC content between older and younger adults (80% power, alpha = 0.05). With respect to serum vitamin D_3, using published data [17,18], we estimated n = 10 participants in each age group would give 80% power to detect a mean difference between younger and older groups of approximately 17 nmol/L and 12 nmol/L at 24 h and 7 days post-UVR, respectively, at a 5% significance level.

The 7DHC data were transformed using the y = 1/y equation, after which the data were normally distributed and parametric tests were performed. Vitamin D_3 results were assessed for normality using the Shapiro–Wilk test and QQ plots (data were normally distributed); similarly, 25(OH)D_3 data were assessed for normality using the Shapiro–Wilk test and QQ plots. Data were analysed using mixed-effects analysis and the multiple comparisons (Sidak correction) test using GraphPad Prism statistical software (version 8.4.3, 10 June 2020).

3. Results

3.1. Volunteers

Twenty-five volunteers were recruited, with 14 and 11 in younger and older groups, respectively. Post-recruitment, three younger participants were excluded (one withdrew participation, one took a sunny holiday, and one had an abnormally high PTH concentration (13.1 pmol/L); see Figure 2, CONSORT patient flow diagram).

One older volunteer was excluded due to total 25(OH)D being > 2 SD above the mean of the cohort (suggesting the volunteer was taking a higher dose of vitamin D supplements than allowed per study protocol). In total, 11 younger adults and 10 older adults were included in the analysis (Table 1). The 7DHC data were excluded for two younger volunteers in their 24 h post-UVR biopsies due to an error that occurred in the protein assay. One older participant was not able to attend the photobiology unit at 24 h post-UVR, such that no biopsies were collected at this timepoint. Routine biochemistry and PTH results were found to be in the normal range for all the included volunteers.

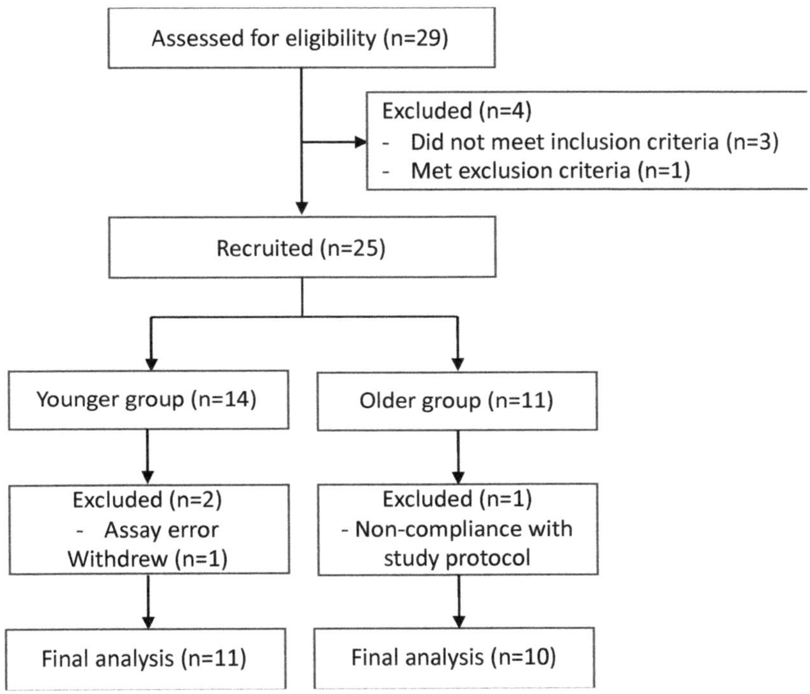

Figure 2. CONSORT patient flow diagram.

Table 1. Volunteer demographics *.

	Younger (18–40 Years Old)	Older (65–89 Years Old)
Participants (n)	11	10
Sex (n): male, female	7, 4	6, 4
Age (years)	29.5 ± 6.3	70.2 ± 3.8
BMI [kg/m^2]	27.1 ± 4.8	26.8 ± 5.6
Skin type I, II, III (n)	3, 2, 6	1, 2, 7

* Mean ± SD values unless otherwise stated. Healthy BMI range is 18.5–25 kg/m^2.

3.2. Skin 7DHC

Baseline 7DHC was very similar, and not lower, in the older group versus the younger group (Table 2, Figure 3). No difference was found in 7DHC concentration between younger and older groups at the unexposed/exposed sites in a mixed-effects analysis ($p = 0.17$; F (2,35) = 1.87). The Sidak test-corrected multiple comparisons showed no difference between groups in unexposed ($p = 0.66$), immediately post-UVR ($p = 0.58$), and 24 h post-UVR ($p = 0.99$) skin.

Table 2. Mean (±SD) 7DHC, serum vitamin D_3, and 25(OH)D_3 concentrations in combined, younger, and older groups of volunteers.

	Groups Combined	Younger	Older
		7DHC (µg/mg)	
Unexposed	0.24 ± 0.08	0.22 ± 0.07	0.25 ± 0.08
Immediately post-UVR	0.25 ± 0.09	0.27 ± 0.10	0.22 ± 0.08
24 h post-UVR	0.27 ± 0.11	0.27 ± 0.08	0.28 ± 0.13
		Serum Vitamin D_3 (nmol/L)	
Baseline	1.5 ± 1.5	1.5 ± 1.5	1.5 ± 1.7
24 h post-UVR	2.8 ± 2.0 [1]	3.1 ± 2.0 [1]	2.5 ± 2.0 [1]
7 d post-UVR	1.8 ± 1.7	2.0 ± 2.1	1.7 ± 1.2
		25(OH)D_3 (nmol/L)	
Baseline	44.2 ± 22.9	37.4 ± 18.9	51.7 ± 25.5
24 h post-UVR	47.7 ± 22.3 [2]	40.5 ± 18.6	54.1 ± 24.3
7 d post-UVR	46.8 ± 21.0	41.3 ± 19.0	52.7 ± 22.4

[1] Significant/near-significant value when compared to baseline in this group; p = 0.0002, p = 0.006, p = 0.06.
[2] Significant value when compared to baseline in this group; p = 0.0445.

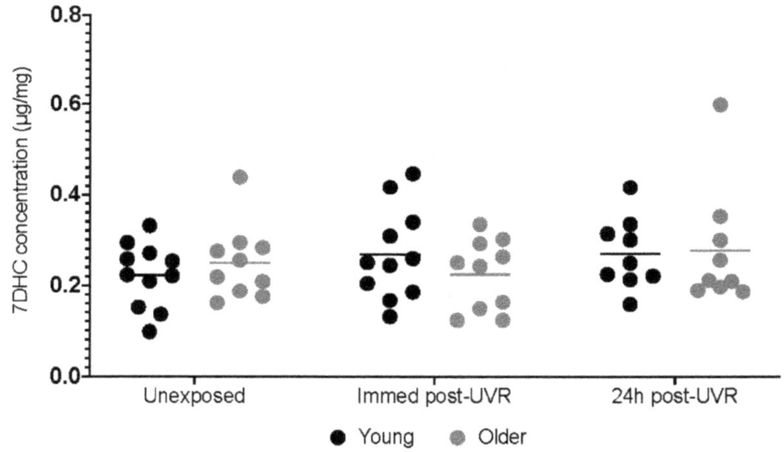

Figure 3. Concentration of skin 7DHC in younger (n = 11/11/9) and older (n = 10/10/9) adults at baseline immediately post-UVR and 24 h post-UVR, respectively. Statistical analyses were performed on transformed data (y = 1/y). Horizontal lines indicate means.

3.3. *Serum Vitamin D_3*

As the limit of detection of the vitamin D_3 assay was <1 nmol/L, all values below this level were treated as 0.5 nmol/L for data analysis purposes (Table 2, Figure 4). A mixed-effect analysis showed that time was a significant factor influencing vitamin D_3 concentrations in both age groups ($p = 0.002$; F (1.645, 27.97) = 8.462), indicating that the single sub-erythemal UVR dose influenced vitamin D_3 concentration. Age and interaction of time and age were not statistically significant as determinants of vitamin D_3 level. At baseline, there was no difference in serum vitamin D_3 between the age groups (the mean in both groups was 1.5 nmol/L). The increase from baseline to 24 h post-UVR was 107% and 67% in younger and older groups, respectively ($p = 0.006$ and $p = 0.06$, respectively, multiple comparisons analysis). In both groups, serum vitamin D_3 decreased by 7 days, and no differences were found between baseline level and that at 7 days post-UVR.

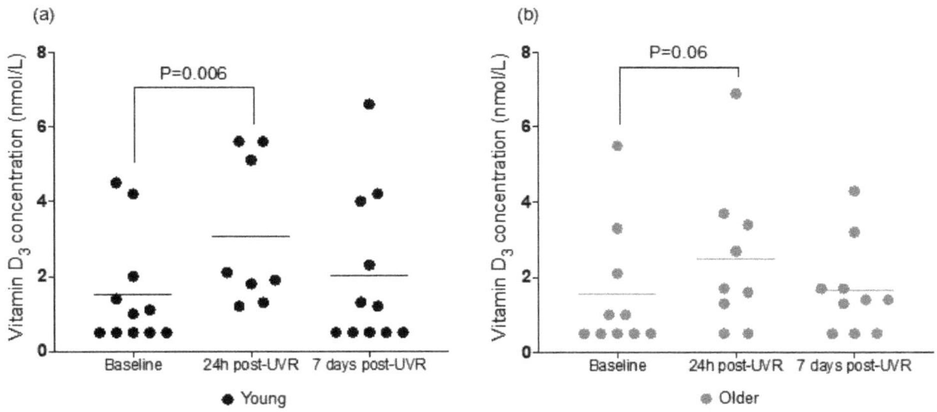

Figure 4. Serum vitamin D_3 concentration in (**a**) younger (n = 11/8/11) and (**b**) older (n = 10/9/10) adults at baseline, 24 h post-UVR, and 7 days post-UVR. Horizontal lines indicate means.

3.4. Serum 25(OH)D$_3$

Serum 25(OH)D_3 and 25(OH)D_2 were assayed alongside vitamin D_3 as the standard measure of vitamin D status. We report on 25(OH)D_3, the only form of 25(OH)D obtained from skin production (Table 2). Mean 25(OH)D_3 concentrations were not lower in the older group than the younger group, contrary to traditional expectations [1,19], potentially reflecting their health and ambulance. Mixed-effects analysis showed no difference in 25(OH)D_3 in the groups combined at different timepoints (F (0.1800, 3.240) = 1.416; $p = 0.16$), while multiple comparison testing showed an increase in 25(OH)D_3 between the baseline and the 24 h post-UVR level ($p = 0.045$). No difference was found in age, time, or interaction of time and age for 25(OH)D_3 in the younger versus the older groups.

4. Discussion

This study was performed to explore whether the cutaneous portion of the vitamin D_3 biosynthetic pathway differs between younger and older adults. Under standardised conditions, and following our development of an updated skin 7DHC assay [12], we showed that skin 7DHC concentration is not lower in older adults (mean 0.25 μg/mL) than younger adults (0.22 μg/mL), contrasting with a previous study of influence of age on 7DHC [10]. We additionally examined the impact of low-dose SSR in vivo on skin 7DHC and serum vitamin D_3, i.e., the metabolite bridging the skin and the circulation in the vitamin D biosynthetic pathway. No detectable change occurred in 7DHC concentration of biopsies taken at two timepoints following UVR exposure, potentially reflecting the small percentage of 7DHC converted to pre-vitamin D, while serum vitamin D_3 increased at 24 h following UVR exposure in both groups. Consistent with this, there was a significant increase in vitamin D status (25(OH)D_3) from baseline to 24 h following UVR in the combined groups. Our findings indicate that baseline skin 7DHC concentration is not a limiting factor for vitamin D_3 synthesis in healthy older adults relative to younger adults, and that response to low-dose UVR is similar in older and younger adults.

The data collected in the current study indicated that skin 7DHC concentration was very similar in healthy volunteers in younger and older groups at baseline, and indeed under all conditions examined (unexposed skin, skin immediately post-UVR, and at 24 h post-UVR). This is in contrast with data reported by MacLaughlin and Holick in 1985, where an approximately two-fold lower baseline epidermal 7DHC content was found in surgical patients 70 years of age versus 30 years of age [10]. Their findings could be attributable to the widely differing skin sites sampled between patients and age groups, and also potentially to the suboptimal health status of some tissue [10,16]. Additionally, the skin

7DHC content was normalised using the surface area and wet weight of samples, which can be unreliable due to potential water loss by evaporation; normalisation by dry weight [20] or protein content [21,22], as we have used, is a more consistent method. In a technical study evaluating a skin 7DHC quantification method [11], a seven-fold variation was found in the 7DHC concentration (n = 25, 12.1–80.6 ug/g dry weight), while no correlation was seen with age. Additionally, both of the earlier studies used an LC–UV system to quantify the 7DHC, which can be challenging when compounds of similar molecular weights are found in the extracted mixture.

The amount of 7DHC that is converted to pre-vitamin D_3 after one UVR exposure appears to be relatively small based on in vitro and ex vivo studies [23]. Thus, it was found that within 5 min of summer sunlight exposure (Boston, MA, USA, 42° N), 3% of 7DHC in methanol in vitro was converted to pre-vitamin D_3, and after 1 h, ~9% was converted. Inert isomers were formed from pre-vitamin D_3, which accumulated with time. Similarly, in ex vivo skin experiments, after 3 h of sunlight exposure, only 7% of 7DHC was converted to pre-vitamin D_3 [23]. In skin specimens irradiated ex vivo with narrowband UVB (295 $\pm$ 5 nm; 0.15 J/cm^2), the conversion rate of 7DHC to pre-vitamin D_3 ranged from 8 to 17%, with the highest rates in the youngest patients, and at a rate of 24 to 33% if evaluating the epidermis separately; these higher conversion rates may reflect the artificially narrow UV waveband used [10]. MacLaughlin and Holick [10] also irradiated skin specimens from five patients ex vivo with 0.15 J/cm^2 UVB (295 $\pm$ 5 nm) and found that the production of pre-vitamin D from 7DHC in very young people (aged 8 years and 18 years) was two-fold higher than in the three older patients (aged 77 years, 77 years, and 82 years), with mean conversions of 31.9% and 25.8%, respectively. In contrast, our protocol involved UVR exposure of emission spectrum (UVB and UVA) and a dose mimicking ~15 min of midday summer sunlight exposure, and our in vivo approach allowed for the potential influence of blood flow, body temperature, and neighbouring tissues; this improved representation of biology may have contributed to any reduction in 7DHC being too small for detection, and/or the 7DHC may have been rapidly restored. Additionally, the number of post-UVR timepoints that could be subjected to skin biopsy was limited by ethical acceptability.

Changes in circulating vitamin D_3 following UVR exposure also reflect the early stage of vitamin D metabolism. There was no difference in the mean baseline value of circulating vitamin D_3 between age groups (both were 1.5 nmol/L), and vitamin D_3 was observed to increase in both age groups at 24 h after UVR, indicating the conversion of 7DHC to pre-vitamin D_3. Although vitamin D_3 increased by 107% and 67% in the younger and older groups, respectively, there was no statistically significant difference between groups. At present, there are no comparable published studies examining the influence of age on circulating vitamin D_3 response to UVR exposure. However, studies using a variety of protocols have examined the impact of UVR on the level of serum vitamin D_3, and our baseline measurements are similar to those of other recent studies reporting average levels of ~1.3 and 2.5 nmoL/L [18]. Other studies generally show larger average increases in serum vitamin D_3 post-UVR [17], which can be accounted for by protocol differences, including higher UVR dose, higher UVB content (converts 7DHC to pre-vitamin D_3) relative to UVA (in contrast, leads to production of inactive isomers), repeated exposures, and more BSA exposed. Thus, reported average increases in serum vitamin D_3 post-UVR have ranged from 6.5 nmoL/L in 15 adults aged 23 $\pm$ 2.3 years given a single UVR dose of 2.2 $\pm$ 0.8 SED to ~50% BSA [18], to 37.7 nmol/L in 10 volunteers aged 48 $\pm$ 12 years given 0.66–1.4 SED on 3 consecutive days to ~50% BSA [17]. However, similarly to Libon et al. [18], who measured serum vitamin D_3 concentration 5 days post-UVR, the serum vitamin D_3 concentration in our study decreased towards baseline at 7 days post-UVR.

The strengths of our study include its prospective design, use of healthy volunteers, standardised procedures and skin sites, mimicking of casual natural sunlight exposure, use of modern assays, and originality of the approach. We recruited volunteers of matched skin type, and skin samples were collected from a body site protected from the sun.

Furthermore, the study was conducted during winter months in the UK, when ambient UVB is insufficient to produce appreciable vitamin D_3 in skin. The UV irradiation protocol was designed to mimic natural sunlight exposure in humans, using UVR comprising 95% and 5% UVB (similar to ambient UVR in summer), approximately 35% BSA exposure (as when wearing summer clothes, such as T-shirt and shorts), and a low sub-erythemal UVR dose that is believed to be the most efficient in vitamin D_3 synthesis [9]. We previously showed that such low doses of SSR can increase vitamin D status whilst minimising but not eliminating the risk of skin damage in lighter skin types [24,25].

A limitation of the study is the relatively low number of volunteers, as intensive in vivo multiple biopsy studies are challenging to recruit and perform in higher numbers, and this potentially resulted in missed detection of significant differences. Repeated UVR doses might also be needed to detect a difference in 7DHC concentration following UVR exposure. Similarly, repeated low doses of UVR, as well as a study with a larger volunteer sample size, are indicated to further explore serum vitamin D_3 differences between age groups post-UVR exposure.

Taken together with inconsistent evidence of the benefits of vitamin D supplements for musculoskeletal health in vitamin D-replete healthy older adults [26], our study may influence assumptions regarding oral vitamin D requirements [27] in the >65-year-old age group. Our findings do not negate the need to be mindful of vitamin D requirements in older adults in residential care or with lower levels of independent mobility, as they can have significantly reduced sunlight exposure and hence lower vitamin D status [28].

5. Conclusions

For approximately 40 years, it has been suggested that vitamin D nutrition is compromised in older adults at least partially through lower levels of the skin precursor 7DHC. However, data from our prospective, standardised study indicate that baseline 7DHC concentration is not lower in healthy older adults compared to younger adults. Similar synthesis of vitamin D_3 was seen in the different age groups upon exposure to a low, sub-sunburn dose of UVR, as can be gained upon casual exposure in daily life. This study provides information assisting health guidance on sun exposure and vitamin D nutrition, which is particularly relevant for the growing populations in many countries of healthy ambulant adults aged >65 years.

Author Contributions: Conceptualization, L.E.R. and A.R.W.; methodology, L.E.R., A.R.W. and W.D.F.; validation, J.J.D. and J.C.Y.T.; formal analysis, O.B., J.J.D. and J.C.Y.T.; investigation, O.B., J.J.D., J.C.Y.T., L.E.R., A.R.W. and W.D.F.; data curation, O.B., J.J.D. and J.C.Y.T.; writing—original draft preparation, O.B.; writing—review and editing, L.E.R., A.R.W., W.D.F., J.C.Y.T. and J.J.D.; supervision, L.E.R., A.R.W. and W.D.F.; project administration, A.R.W. and L.E.R.; funding acquisition, A.R.W., L.E.R. and W.D.F. All authors have read and agreed to the published version of the manuscript.

Funding: The research was funded by the Biotechnology and Biological Sciences Research Council (BBSRC) Doctoral Training Programme (DTP) (student number 7617061) and supported by the NIHR Manchester Biomedical Research Centre (NIHR203308). The views expressed are those of the author(s) and not necessarily those of the NIHR or the Department of Health and Social Care.

Institutional Review Board Statement: The North West Greater Manchester West Research Ethics Committee, aligning with the Declaration of Helsinki reference 18/NW/0493; 25 July 2018.

Informed Consent Statement: Written informed consent was obtained from all study participants.

Data Availability Statement: Data required to assess the manuscript's conclusions are provided within the manuscript. Further data relevant to this research can be found on the ISRCTN trial registry website (ISRCTN no. 72674753).

Conflicts of Interest: None relevant. A.R.W. acknowledges non-personal funding for unrelated UV equipment characterisation from Signify, Philips (the Netherlands), and for unrelated vitamin D advice from the Walgreen Boots Alliance (UK).

References

1. Boucher, B.J. The Problems of Vitamin D Insufficiency in Older People. *Aging Dis.* **2012**, *3*, 313–329.
2. Christensen, K.; Doblhammer, G.; Rau, R.; Vaupel, J.W. Ageing Populations: The Challenges Ahead. *Lancet.* **2009**, *374*, 1196–120 [CrossRef]
3. Office for National Statistics, Census: Population Estimates for the United Kingdom. 2011. Available online: https://www.on gov.uk/peoplepopulationandcommunity/populationandmigration/populationestimates (accessed on 28 January 2022).
4. Ashwell, M.; Stone, E.M.; Stolte, H.; Cashman, K.D.; Macdonald, H.; Lanham-New, S.; Him, S.; Webb, A.; Fraser, D. UK Foo Standards Agency Workshop Report: An Investigation of the Relative Contributions of Diet and Sunlight to Vitamin D Status. *B J. Nutr.* **2010**, *104*, 603–611. [CrossRef]
5. Nemanic, M.K.; Whitney, J.; Arnaud, S.; Herbert, S.; Elian, P.M. Vitamin D_3 Production by Cultured Human Keratinocytes an Fibroblasts. *Biochem. Biophys. Res. Commun.* **1983**, *115*, 444–450. [CrossRef]
6. Slominski, A.; Zjawiony, J.; Wortsman, J.; Semak, I.; Stewart, J.; Pisarchik, A.; Sweatman, T.; Marcos, J.; Dunbar, C.; Tuckey, R.C. Novel Pathway for Sequential Transformation of 7-Dehydrocholesterol and Expression of the P450scc System in Mammalia Skin. *Eur. J. Biochem.* **2004**, *271*, 4178–4188. [CrossRef]
7. CIE. Action Spectrum for the Production of Previtamin D3 in Human Skin; Technical Report 174. 2006. Available onlin https://cie.co.at/publications/action-spectrum-production-previtamin-d3-human-skin (accessed on 16 February 2019).
8. Wacker, M.; Holick, M.F. Sunlight and Vitamin D. *Dermatoendocrinol* **2013**, *5*, 51–108. [CrossRef]
9. Webb, A.R.; Decosta, B.R.; Holick, M.F. Sunlight Regulates the Cutaneous Production of Vitamin D3 by Causing Its Photodegr dation. *J. Clin. Endocrinol. Metab.* **1989**, *68*, 882–887. [CrossRef]
10. MacLaughlin, J.; Holick, M.F. Aging Decreases the Capacity of Human Skin to Produce Vitamin D_3. *J. Clin. Investig.* **1985**, *7* 1536–1538. [CrossRef]
11. Moody, J.P.; Humphries, C.A.; Allan, S.M.; Paterson, C.R. Determination of 7-Dehydrocholesterol in Human Skin by Hig Performance Liquid Chromatography. *J. Chromatogr.* **1990**, *530*, 19–27. [CrossRef]
12. Borecka, O.; Rhodes, L.E.; Webb, A.R.; Dutton, J.J.; Fraser, W.D. A Newly Developed and Validated LC-MS/MS Method fo Measuring 7-dehydrocholesterol (7DHC) Concentration in Human Skin: A Tool for Vitamin D Photobiology Research. *Photochem Photobiol. Sci.* **2022**, *21*, 2001–2009. [CrossRef]
13. Tang, J.C.Y.; Nicholls, H.; Piec, I.; Washbourne, C.J.; Dutton, J.J.; Jackson, S.; Greeves, J.; Fraser, W.D. Reference Intervals for Serur 24,25-Dihydroxyvitamin D and the Ratio with 25-Hydroxyvitamin D Established using a Newly Developed LC-MS/MS Metho *J. Nutr. Biochem.* **2017**, *46*, 21–29. [CrossRef]
14. Fitzpatrick, T.B. The Validity and Practicality of Sun-Reactive Skin Types I through VI. *Arch. Dermatol.* **1988**, *124*, 869. [CrossRef] [PubMed]
15. Bio-Rad. Bio-Rad Protein Assay. Available online: https://www.bio-rad.com/en-uk/product/bio-rad-protein-assay?ID=d4d41 9a-12e8-4819-8b3e-ccab019c6e13 (accessed on 28 May 2022).
16. SACN. Vitamin D and Health. Available online: https://www.gov.uk/government/uploads/system/uploads/attachment data/file/537616/SACN_Vitamin_D_and_Health_report.pdf (accessed on 28 January 2022).
17. Osmancevic, A.; Sandström, K.; Gillstedt, M.; Landin-Wilhelmsen, K.; Larkö, O.; Wennberg Larkö, A.M.; Holick, M.F.; Krogsta A.L. Vitamin D Production after UVB Exposure—A Comparison of Exposed Skin Regions. *J. Photochem. Photobiol. B* **2015**, *14* 38–43. [CrossRef] [PubMed]
18. Libon, F.; Courtois, J.; Goff, C.L.; Lukas, P.; Fabregat-Cabello, N.; Seidel, L.; Cavalier, E.; Nikkels, A.F. Effect of Body Site an Surface on Vitamin D and 25-Hydroxyvitamin D Production after a Single Narrowband UVB Exposure. *J. Investig. Dermatol.* **201** *137*, 1391–1393. [CrossRef]
19. Meehan, M.; Penckofer, S. The Role of Vitamin D in the Aging Adult. *J. Aging Gerontol.* **2014**, *2*, 60–71. [CrossRef]
20. Coger, V.; Million, N.; Rehbock, C.; Sures, B.; Nachev, M.; Barcikowski, S.; Wistuba, N.; Strauß, S.; Vogt, P.M. Tissue Concentration of Zinc, Iron, Copper, and Magnesium During the Phases of Full Thickness Wound Healing in a Rodent Model. *Biol. Trace Elem Res.* **2019**, *191*, 167–176. [CrossRef] [PubMed]
21. Kendall, A.C.; Pilkington, S.M.; Sassano, G.; Rhodes, L.E.; Nicolaou, A. N-Acyl Ethanolamide and Eicosanoid Involvement i Irritant Dermatitis. *Br. J. Dermatol.* **2016**, *175*, 163–171. [CrossRef]
22. Liu, W.; Xu, L.; Lamberson, C.; Haas, D.; Korade, Z.; Porter, N.A. A Highly Sensitive Method for Analysis of 7-Dehydrocholesterc for the Study of Smith-Lemli-Opitz Syndrome. *J. Lipid Res.* **2014**, *55*, 329–337. [CrossRef]
23. Webb, A.R.; Kline, L.; Holick, M.F. Influence of Season and Latitude on the Cutaneous Synthesis of Vitamin D_3: Exposure t Winter Sunlight in Boston and Edmonton Will Not Promote Vitamin D_3 Synthesis in Human Skin. *J. Clin. Endocrinol. Metab.* **198** *67*, 373–378. [CrossRef]
24. Rhodes, L.E.; Webb, A.R.; Fraser, H.I.; Kift, R.; Durkin, M.T.; Allan, D.; O'Brien, S.J.; Vail, A.; Berry, J.L. Recommended Summe Sunlight Exposure Levels Can Produce Sufficient ($\geq$20 ngmL^{-1}) but Not the Proposed Optimal ($\geq$32 ngmL^{-1}) 25(OH)D Level at UK Latitudes. *J. Investig. Dermatol.* **2010**, *130*, 1411–1418. [CrossRef]
25. Felton, S.J.; Cooke, M.S.; Kift, R.; Berry, J.L.; Webb, A.R.; Lam, P.M.; de Gruijl, F.R.; Vail, A.; Rhodes, L.E. Concurrent beneficia (vitamin D production) and hazardous (cutaneous DNA damage) impact of repeated low-level summer sunlight exposures. *Br. Dermatol.* **2016**, *175*, 1320–1328. [CrossRef] [PubMed]

26. LeBoff, M.S.; Bischoff-Ferrari, H.A. The Effects of Vitamin D Supplementation on Musculoskeletal Health: The VITAL and DO-Health Trials. *J. Gerontol. A Biol. Sci. Med.* **2023**, *78*, 73–78. [CrossRef] [PubMed]
27. Public Health England: Scientific Advisory Committee on Nutrition. Vitamin D and Health. 2016. Available online: https://www.gov.uk/government/publications/sacn-vitamin-d-and-health-report (accessed on 5 February 2019).
28. Webb, A.R.; Pilbeam, C.; Hanafin, N.; Holick, M.F. A long-term study to evaluate the relative contributions of exposure to sunlight and diet on the circulating concentrations of 25-OH-D in an elderly population in Boston. *Am. J. Clin. Nutr.* **1990**, *51*, 1075–1081. [CrossRef] [PubMed]

Disclaimer/Publisher's Note: The statements, opinions and data contained in all publications are solely those of the individual author(s) and contributor(s) and not of MDPI and/or the editor(s). MDPI and/or the editor(s) disclaim responsibility for any injury to people or property resulting from any ideas, methods, instructions or products referred to in the content.

nutrients

MDPI

Review

Vitamin D and Risk of Incident Type 2 Diabetes in Older Adults: An Updated Systematic Review and Meta-Analysis

Ligia J. Dominguez [1,*,†], Nicola Veronese [2,†], Eliana Marrone [2], Carla Di Palermo [2], Candela Iommi [2], Rosaria Ruggirello [2], Carla Caffarelli [3], Stefano Gonnelli [3] and Mario Barbagallo [2]

1. Department of Medicine and Surgery, "Kore" University of Enna, 94100 Enna, Italy
2. Geriatric Unit, Department of Medicine, University of Palermo, 90127 Palermo, Italy; nicola.veronese@unipa.it (N.V.); eliana94marrone27@gmail.com (E.M.); carladipalermo@gmail.com (C.D.P.); candelaiommi@gmail.com (C.I.); rosariaruggirello34@gmail.com (R.R.); mario.barbagallo@unipa.it (M.B.)
3. Section of Internal Medicine, Department of Medicine, Surgery and Neuroscience, University of Siena, 53100 Siena, Italy; carla.caffarelli@unisi.it (C.C.); stefano.gonnelli@unisi.it (S.G.)

* Correspondence: ligia.dominguez@unikore.it

† These authors equally contributed to this work.

Abstract: Vitamin D deficiency is very common worldwide, particularly in old age, when people are at the highest risk of the negative adverse consequences of hypovitaminosis D. Additionally to the recognized functions in the regulation of calcium absorption, bone remodeling, and bone growth, vitamin D plays a key role as a hormone, which is supported by various enzymatic, physiological, metabolic, and pathophysiological processes related to various human organs and systems. Accruing evidence supports that vitamin D plays a key role in pancreatic islet dysfunction and insulin resistance in type 2 diabetes. From an epidemiological viewpoint, numerous studies suggest that the growing incidence of type 2 diabetes in humans may be linked to the global trend of prevalent vitamin D insufficiency. In the past, this association has raised discussions due to the equivocal results, which lately have been more convincing of the true role of vitamin D supplementation in the prevention of incident type 2 diabetes. Most meta-analyses evaluating this role have been conducted in adults or young older persons (50–60 years old), with only one focusing on older populations, even if this is the population at greater risk of both hypovitaminosis D and type 2 diabetes. Therefore, we conducted an update of the previous systematic review and meta-analysis examining whether hypovitaminosis D (low serum 25OHD levels) can predict incident diabetes in prospective longitudinal studies among older adults. We found that low 25OHD was associated with incident diabetes in older adults even after adjusting for several relevant potential confounders, confirming and updating the results of the only previous meta-analysis conducted in 2017.

Keywords: vitamin D; diabetes; hypovitaminosis D; aging; elderly; older adults; meta-analysis

1. Introduction

According to the International Diabetes Federation (IDF) Diabetes Atlas, the global diabetes prevalence in people 20–79 years old in 2021 was estimated to be 10.5% (536.6 million people), with a prospective 12.2% increase (783.2 million) by 2045. The number of people diagnosed with type 2 diabetes has increased fivefold in the past 25 years. As in previous estimates, diabetes prevalence was highest in older adults, particularly in those aged 75–79 years. Due to its association with relevant complications, global diabetes-related health expenditures were estimated at 966 billion USD in 2021 and are projected to reach 1054 billion USD by 2045 [1]. These current numbers and projections make diabetes one of the most devastating diseases of the 21st century [2].

Long-lasting diabetes can lead to a wide range of complications such as chronic pain [3], lower limb amputations [4], increased healthcare costs [5–7], falls [8], and cognitive decline and dementia [9,10], among others. Even if pharmacological treatments for diabetes have recently advanced significantly, there are still no effective interventions in the prevention of this alarming disease, and governments have proven to be slow to adopt the action necessary for diabetes prevention [11].

A deficit of vitamin D is a potential reversible target for diabetes prevention in older adults. In fact, another condition for which interest has grown in recent years is hypovitaminosis D, which has been increasingly reported to be associated with several negative health outcomes, including an increasing risk of incident diabetes [12,13]. Furthermore, vitamin D deficit is extremely frequent, particularly in old age [14] as well as in patients with type 2 diabetes [15].

Several mechanisms may help explain the link between vitamin D and type 2 diabetes, including vitamin D-mediated stimulation of pancreatic insulin secretion, the association of hypovitaminosis D with metabolic syndrome and with inflammation, as well as genetic polymorphisms of vitamin D that may lead to impaired blood glucose control [16,17].

Previous observational studies showed an inverse association between 25 hydroxyvitamin D (25OHD) and the risk of incident type 2 diabetes confirmed in a meta-analysis by Song et al., including twenty-one prospective observational studies [18]. However, intervention studies have been either negative or showed only limited benefit [19,20]. A former meta-analysis of eight trials did not show any effect of supplementing vitamin D on incident diabetes or glycemic control [21]. More recently, a meta-analysis by Barbarawi et al., including nine randomized controlled trials (RCTs) (43,559 participants), reported that in patients with prediabetes, vitamin D supplementation at moderate to high doses (≥1000 IU/day) significantly reduced the risk of incident type 2 diabetes, compared with the placebo [22]. However, the significant association persisted only for those with a body mass index (BMI) lower than 30 kg/m^2 and was not significant for those with a BMI higher than 30 kg/m^2. Similarly, a meta-analysis by Zhang et al., including eight RCTs of patients with prediabetes (4896 participants), comparing vitamin D vs. the placebo (regardless of dose, type, route of administration, and duration) showed that overall vitamin D supplementation significantly reduced the risk of type 2 diabetes, but group analysis showed that the benefit was found in nonobese subjects but not in obese subjects [23]. Pittas et al. have recently reported the results of a meta-analysis including only RCTs specifically designed and conducted to test the effects of oral vitamin D vs. the placebo on new-onset diabetes in adults with prediabetes [24]. Among three RCTs with a low risk of bias, vitamin D reduced the risk for incident diabetes by 15% in adjusted analyses. In participants in the intervention arm who maintained a mean serum 25OHD level ≥50 ng/mL vs. 20–29 ng/mL throughout follow-up, vitamin D supplementation reduced the risk for diabetes by 76%, with a 3-year absolute risk reduction of 18.1%. Even if the results are remarkable, the authors admit that they may not be directly applicable to the general population because the studies included people with prediabetes, which was an inclusion criterion of the meta-analysis [24].

All the above-mentioned meta-analyses [18,22–24], which support hypovitaminosis D as a possible key determinant of incident type 2 diabetes, have been done in populations of adults or young older adults (aged 50–60 years). Most studies excluded older people, even if this is the population at greater risk for both hypovitaminosis D [14,25] and type 2 diabetes [1,26]. A previous meta-analysis, including nine prospective studies (28,258 participants), investigated whether hypovitaminosis D was associated with incident diabetes among older participants (mean age: 67.7 years; median follow-up: 7.7 years). The authors found that lower levels of 25OHD were associated with a significantly higher risk of incident diabetes compared with higher levels of 25OHD. The results remained significant even after adjustments for a median of eleven potential confounders among the available studies [27].

Given this background, we conducted an update of the previous systematic review and meta-analysis examining whether hypovitaminosis D (a low serum 25OHD level) can predict incident diabetes in prospective longitudinal studies conducted among older adults.

2. Materials and Methods

The present systematic review was performed following strengthening the reporting of observational studies in epidemiology [STROBE] criteria [28] and the recommendations in the preferred reporting items for systematic reviews and meta-analyses [PRISMA] statement [29].

2.1. Search Strategy

Four authors in couples independently searched for longitudinal studies that considered diabetes and serum 25OHD levels among older people. Major databases (PubMed, SCOPUS) were examined from 10 July 2016 until 15 October 2023, without language restrictions.

The search strategy used in PubMed was: (vitamin D OR 25-hydroxyvitamin D OR calcidiol OR calcitriol OR calcifediol) AND (insulin OR glucose OR beta-cell function OR diabet*) AND (old OR elderly OR older OR aged). An analogous search (adapted to the requirements of Scopus) was performed.

2.2. Eligibility Criteria and Study Selection

We considered the following requisites for articles to be eligible: (1) longitudinal prospective design; (2) assessment of serum 25OHD considering that it is currently accepted as the best indicator for vitamin D status [19]; (3) including only older people (mean age of participants ≥60 years; (4) self-reported diagnosis of diabetes during follow-up, or diagnosis through medical and/or hospital records, or at least one diagnostic criterion according to the American Diabetes Association (i.e., glycosylated hemoglobin (HbA1c) ≥6.5%, fasting plasma glucose (FPG) ≥126 mg/dL, or use of antidiabetic medications). Studies with the following characteristics were excluded: (1) having a cross-sectional design; (2) assessment of vitamin D status with methods different than serum 25OHD (e.g., dietary intake of vitamin D); (3) including only subclinical estimations of diabetes (e.g., changes in FPG). References of the included articles were hand-searched in order to identify additional, potentially pertinent publications. Abstracts from conferences were also examined, and in such cases, we contacted the corresponding authors to obtain the data enabling inclusion.

2.3. Data Extraction

Data extracted from the selected studies were independently recorded into a standardized Microsoft Excel spreadsheet by one author checking as well the quality of the extracted data. Any discrepancy was resolved by agreement with a third expert author. For each study, the following information was extracted: (i) characteristics of the study (e.g., demographics, sample size, country in which the study was conducted); (ii) setting of the study; (iii) number of people developing diabetes during the follow-up period; (iv) number of years of follow-up; (v) levels of serum 25OHD and methods utilized for the assessment of 25OHD; (vi) diagnostic criteria considered for the formulation of diabetes diagnosis; (vii) type and number of covariates included in the multivariable analyses. If the study included >2 categories for serum 25OHD, we considered the lowest quantile as low level and the highest quantile as the reference group.

2.4. Study Quality Assessment

One investigator assessed the quality of the study, which was independently checked by another one. The assessment of the study quality was performed using the Newcastle–Ottawa Scale (NOS) [30]. This scale allocates a maximum of nine points according to three quality parameters: selection, comparability, and outcome.

2.5. Statistical Analysis

A random effects meta-analysis was undertaken using MedCalc® Statistical Software version 22.023 (MedCalc Software Ltd., Ostend, Belgium; https://www.medcalc.org; accessed on 1 March 2024) to take into account the expected heterogeneity. The primary analyses synthesize data by calculating pooled RRs and 95% CI (including the number of incident cases of diabetes in higher levels of serum 25OHD vs. lower values). Secondary analyses considered the hazard ratio/odds ratio (HR/OR) adjusted for the highest number of available covariates for each study examined together using fully adjusted RRs. Heterogeneity among the included studies was evaluated using chi-squared and I-squared statistics, assuming a $p < 0.05$ and a value $\geq$ 50%, respectively, indicative of a significant heterogeneity [31]. When we found significant heterogeneity, and there were $\geq$4 studies available, metaregression analysis was preplanned considering the potential moderators: (1) continent in which the study was performed; (2) study setting (community vs. others); (3) follow-up in years (continuous); (4) diagnostic criteria utilized in the formulation of diabetes diagnosis (self-reported vs. others); and the number of covariates; (5) visual inspection of funnel plots and the Egger bias test were used to assess publication bias [32]. We used the Duval and Tweedie nonparametric trim-and-fill method to examine the potential publication bias when $\geq$3 studies were available. Assuming that the effect sizes of all the studies were normally distributed around the center of a funnel plot, in the event of asymmetries, this procedure adjusts for the potential effect of unpublished (trimmed) studies [33].

3. Results

3.1. Literature Search

As shown in Figure 1, after the update, we identified 2653 nonduplicated, potentially eligible studies. Of them, 37 full texts were examined, and three were eligible for the update of our meta-analysis for 12 works [34–45].

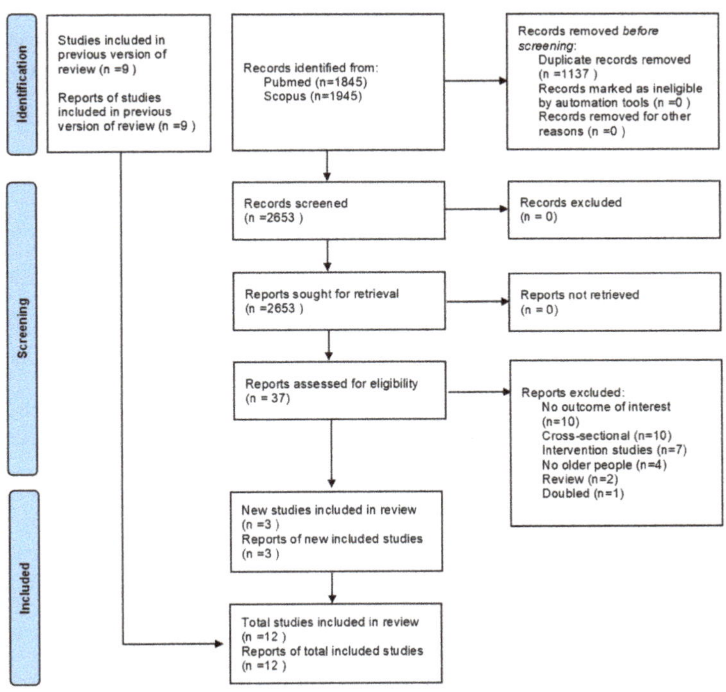

Figure 1. PRISMA flow-chart.

3.2. Study and Participants Characteristics

In Table 1, the characteristics of the 12 included studies involving 40,664 older participants, followed up for a median period of 7.3 (range: 2–14) years, are shown. From the total participants at baseline, 4263 participants (10.5% of the baseline population) developed diabetes. The mean age of the participants was 69.1 years, and most of them were women (66%). All the studies were carried out among community dwellers, mainly in European (n. of studies = 6) and North American (n. of studies = 5) populations. As shown in Table 1, most of the studies used the radioimmunoassay (RIA) method for assessing 25OHD levels. In none of the studies included, did the authors report baseline vitamin D supplementation by the future risk of diabetes. Regarding the diagnosis of diabetes, two studies used only self-reported information. Other studies used self-reported information, followed by a final specialist adjudication in four studies, with a value of FPG over 126 mg/dL in another one and with the use of antidiabetic medications in a third study. Two studies used a value of HbA1c over 6.5% for the diagnosis of type 2 diabetes, one only a value of FPG over 126 mg/dL, and one study considered all the criteria suggested by the American Diabetes Association.

The median NOS values (median = 7; range: 5–9) indicate that the quality of the studies was generally of moderate quality (Table 1).

3.3. Unadjusted Findings Concerning Lower 25OHD Levels and Diabetes

Figure 2 shows the results of the meta-analysis, including 960 older participants with lower 25OHD (from a total of 6717 participants; 14.3%) and 1000 participants with higher 25OHD (from a total of 9207 participants; 10.9%) with a diagnosis of diabetes during the follow-up period. Lower baseline 25OHD levels were associated with a higher risk of incident diabetes vs. older people with higher 25OHD levels: this analysis included a total of 15,924 older participants and reported an RR = 1.20; 95% CI: 1.06–1.35; $p = 0.001$; $I^2 = 29.9\%$). No publication bias was found (Egger's test = 0.26; $p = 0.85$) (Figure 3), and the trim and fill analysis did not modify our results.

Table 1. Description of the included studies characteristics.

Author, Year	Country	Sample Size	N of Subjects with Diabetes at Follow-Up	Mean Age	Percentage of Females	Mean Serum 25OHD Levels	Follow-Up (Years)	Methods of Measurement of Serum 25OHD	Diagnostic Criteria for Diabetes	Type of Covariates	Number of Co-variates	NOS
Bolland A J et al., 2010 [37]	New Zealand	1472	29	74	100	50.9	5	RIA (DiaSorin, Stillwater, MN, USA)	Self-reported with a final adjudication by a specialist	Treatment allocation, age, body weight, smoking status.	4	5
Kositsawat J et al., 2014 [38]	USA	2193	477	74.6	52	65.8	2	RIA (DiaSorin, Stillwater, MN, USA)	Abnormal HbA1c (≥6.5%)	Age, sex, recruitment site, BMI, physical activity, vitamin D supplementation, multivitaminic use, season, PTH, race, diabetes status at follow-up.	11	5
McCarthy, 2022 [35]	Ireland	3373	110	62.9	53.5	57.1	4	Liquid chromatography-mass spectrometry	Self-reported doctor diagnosis of diabetes, use of diabetes medications, or an HbA1c level ≥48 mmol/mol	age, sex, educational status attained, body mass index, smoking history, physical activity, use of statins, and the season	8	7
Napoli N et al., 2016 [39]	USA	1939	139	73.3	0	27.0	6.4	Liquid chromatography tandem mass spectrometry (Thermo Fisher Scientific, Franklin, MA, USA)	FPG ≥ 126 mg/dL, self-reported	Age, site, race, season, BMI, calcium intake	6	7
Pilz S et al., 2012 [40]	The Netherlands	351	45	67.8	61	56.7	7.5	RIA (DiaSorin, Stillwater, MN, USA)	FPG ≥ 126 mg/dL F	Fasting glucose, age, sex, BMI, HDL, TG, hypertension, physical activity level, season, PTH, eGFR	11	7

Table 1. *Cont.*

Author, Year	Country	Sample Size	N of Subjects with Diabetes at Follow-Up	Mean Age	Percentage of Females	Mean Serum 25OHD Levels	Follow-Up (Years)	Methods of Measurement of Serum 25OHD	Diagnostic Criteria for Diabetes	Type of Covariates	Number of Co-variates	NOS
Robinson G et al., 2011 [41]	USA	5140	317	66	100		7.3	DiaSorin Liaison Chemiluminescence Method (DiaSorin, Stillwater, MN)	Self-reported, drugs	Age, ethnicity, latitude, season, WHI study indicators, BMI, hypertension, fiber intake, magnesium intake, physical activity level	10	7
Schafer AL et al., 2014 [42]	USA	5463	320	76.5	100	57.4	8.6	Liquid chromatography–tandem mass spectrometry (Thermo Fisher Scientific, Franklin, MA, USA)	Self-reported	Age, clinic site, BMI, hypertension, self-reported health	5	6
Shottker B et al., 2012 [43]	Germany	7791	829	62	58	46.1	8	RIA (DiaSorin, Stillwater, MN, USA)	HbA1c $\geq$ 6.5%	Age, sex, season, multivitamin supplementation, frequent fish consumption, BMI, HbA1c, family history of diabetes, education, physical activity, smoking, hypertension, renal dysfunction, CRP, fasting TG	15	8

Table 1. *Cont.*

Author, Year	Country	Sample Size	N of Subjects with Diabetes at Follow-Up	Mean Age	Percentage of Females	Mean Serum 25OHD Levels	Follow-Up (Years)	Methods of Measurement of Serum 25OHD	Diagnostic Criteria for Diabetes	Type of Covariates	Number of Co-variates	NOS
Thorand B et al., 2011 [44]	Germany	1683	416	>65	47		11	Enzyme immunoassay IDA, Frankfurt, Germany	Self- reported	Age, sex, survey, season, BMI, smoking, alcohol use, physical activity, systolic BP, total cholesterol/HDL, family history of diabetes, CRP, IL6, ICAM-1, IFN-alpha Inducible protein, markers of inflammation, waist-to-hip ratio.	17	6
Veronese N et al., 2014 [45]	Italy	2227	291	76.1	59	80.1	4.4	RIA (DiaSorin, Stillwater, MN, USA)	FPG $\geq$ 126 mg/dL or HbA1c $\geq$ 6.5% or use of medications	Age, gender, waist, hypertension, education, monthly income, smoking, eGFR, FPG, HbA1c, total cholesterol	11	8
Xia, 2021 [36]	USA	4191	453	65	100	52.0	12	Enzyme immunoassay with a competitive binding technique from Immunodiagnostic Systems Inc. (Fountain Hills, AZ, USA).	self-report of a new physician diagnosis of diabetes treated with oral drugs or insulin shots during study follow-up,	Age, clinical center, race/ethnicity, BMI, family history of diabetes, educational levels, alcohol intake, physical activity levels, cigarette smoking status, postmenopausal hormone therapy use, and season	11	7

Table 1. *Cont.*

Author, Year	Country	Sample Size	N of Subjects with Diabetes at Follow-Up	Mean Age	Percentage of Females	Mean Serum 25OHD Levels	Follow-Up (Years)	Methods of Measurement of Serum 25OHD	Diagnostic Criteria for Diabetes	Type of Covariates	Number of Co-variates	NOS
Zhu, 2022 [34]	Germany	4841	837	61.9	57.5	45.0	14	RIA (DiaSorin, Stillwater, MN, USA)	General practitioner-confirmed patient self-reports	Age; sex; education; smoking and drinking status; vegetable, fruit, and fish consumption; regular intake of multivitamin supplements; body mass index; HbA1c; total cholesterol; high-density lipoprotein cholesterol; triglycerides; C-reactive protein; systolic blood pressure; estimated glomerular filtration rate; family history of diabetes; history of cardiovascular diseases and cancer; antihypertensive medication; lipid-lowering medication; and season of blood draw.	23	7

Abbreviations: 25OHD = 25-Hydroxyvitamin D, RIA = radioimmunoassay, HbA1c = glycated hemoglobin, BMI = body mass index, PTH = parathyroid hormone, FPG = fasting plasma glucose, HDL = high-density lipoproteins, TG = triglyceride, eGFR = glomerular filtration rate, CRP = C-reactive protein, Systolic BP = systolic blood pressure (BP), IL6 = Interleukin 6, ICAM 1 = Intercellular Adhesion Molecule 1, IFN-alpha = interferons-alpha.

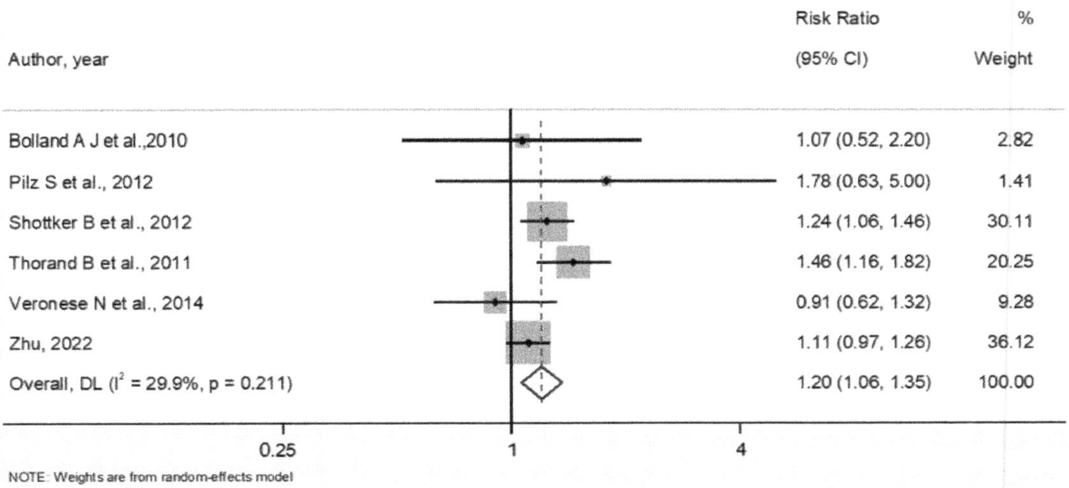

Figure 2. Association between lower 25 hydroxyvitamin D (25OHD) levels at baseline and incident diabetes [34,37,40,43–45].

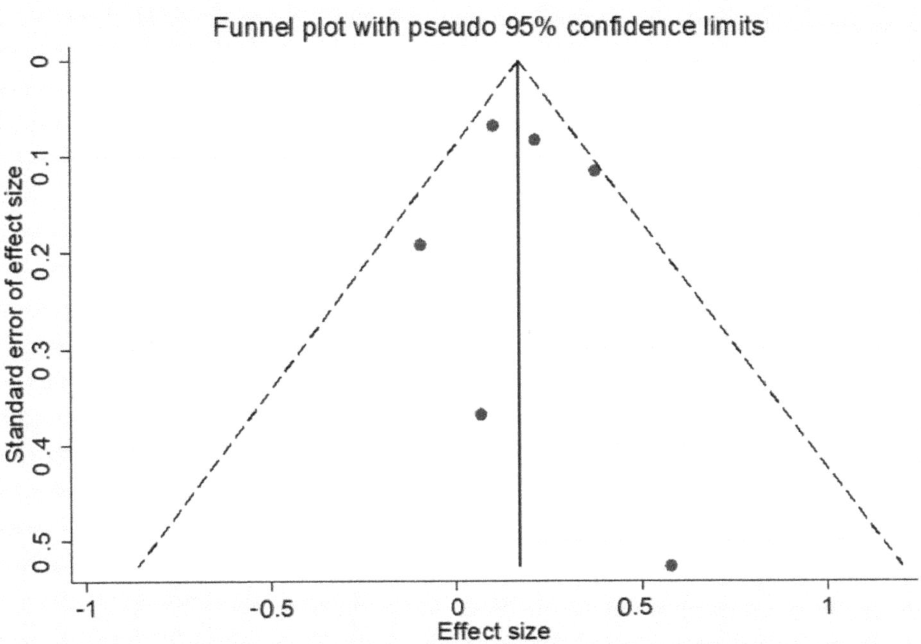

Figure 3. Funnel plot of the unadjusted associations.

3.4. Adjusted Analyses Concerning Lower 25OHD Levels and Diabetes

Figure 4 shows the association between lower baseline serum 25OHD levels and incident diabetes at follow-up after adjusting for potential confounders. The median number of adjustments in the multivariable analyses was 11 (range: 4–17), with all the studies including adiposity estimates, as fully detailed in Table 1. In the 12 studies included, we observed that lower baseline 25OHD levels increased the risk of developing diabetes by 19% (HR = 1.22; 95% CI: 1.12–1.33; $p < 0.0001$; $I^2 = 0\%$). We did not observe publication bias (Egger's test = −0.34; $p = 0.87$) (Figure 5), and the trim and fill analysis did not affect our original findings.

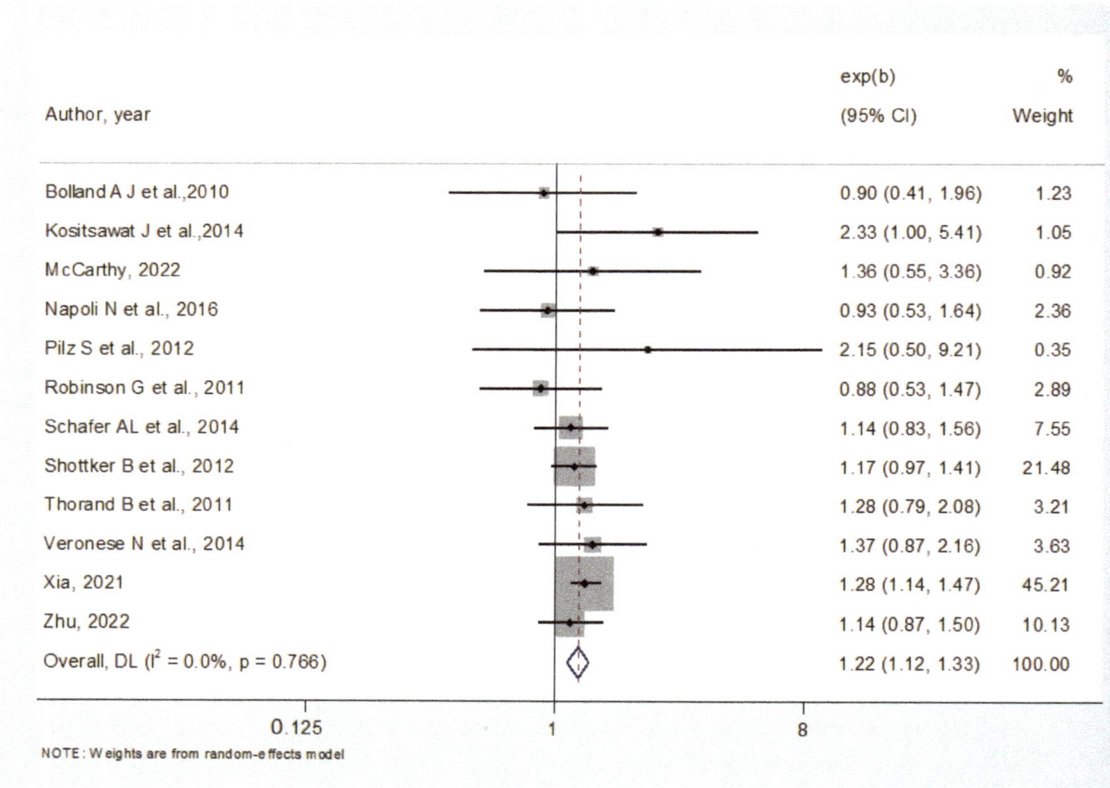

Figure 4. Association between lower 25-hydroxyvitamin D (25OHD) levels at baseline and incident diabetes, after adjusting for potential confounders [34–45].

3.5. Metaregression and Sensitivity Analysis

We did not observe high heterogeneity among the outcomes included in our analyses; therefore, metaregression analyses were not performed. However, since serum 25OHD concentrations may change during the time [46], we did run a metaregression analysis overall, showing that a longer follow-up did not moderate for unadjusted (beta = 0.01; $p = 0.67$; R2 = 0%) or adjusted (beta = 0.03; $p = 0.85$; R2 = 0%) data.

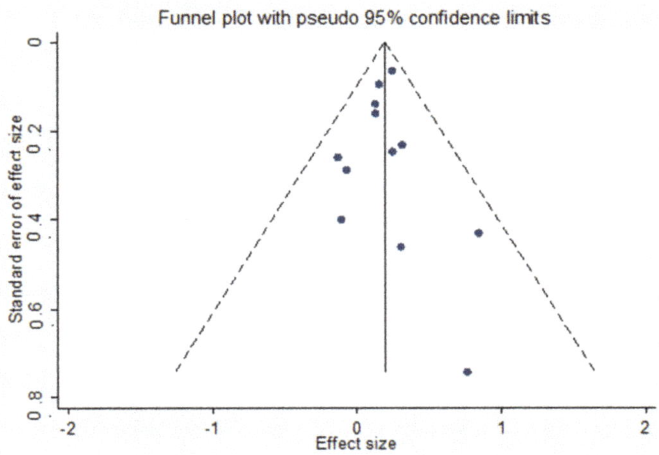

Figure 5. Funnel plot of the adjusted associations.

We also did a cumulative analysis, removing one study at each time, but the results were unchanged, as shown in Figures 6 and 7.

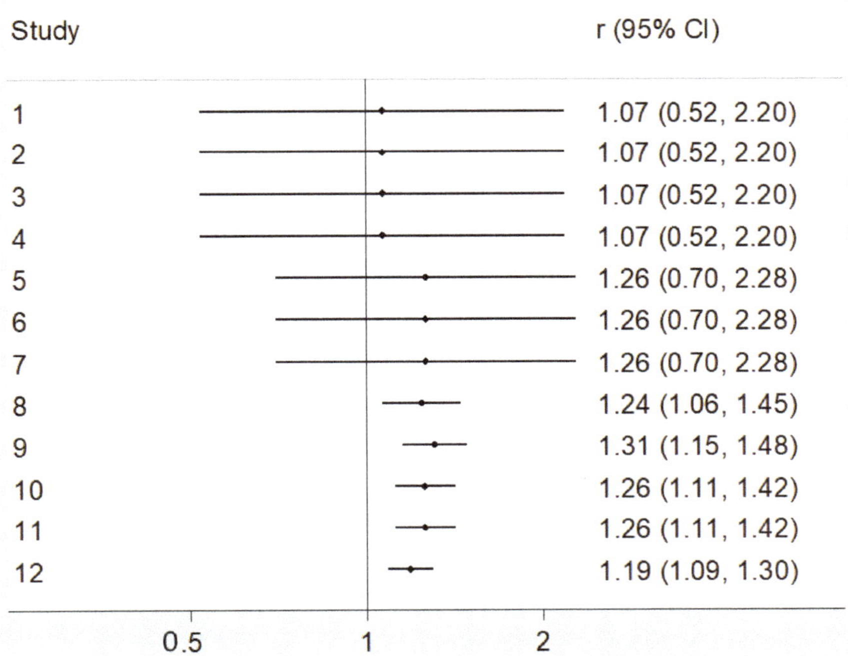

Figure 6. Cumulative analysis for unadjusted data.

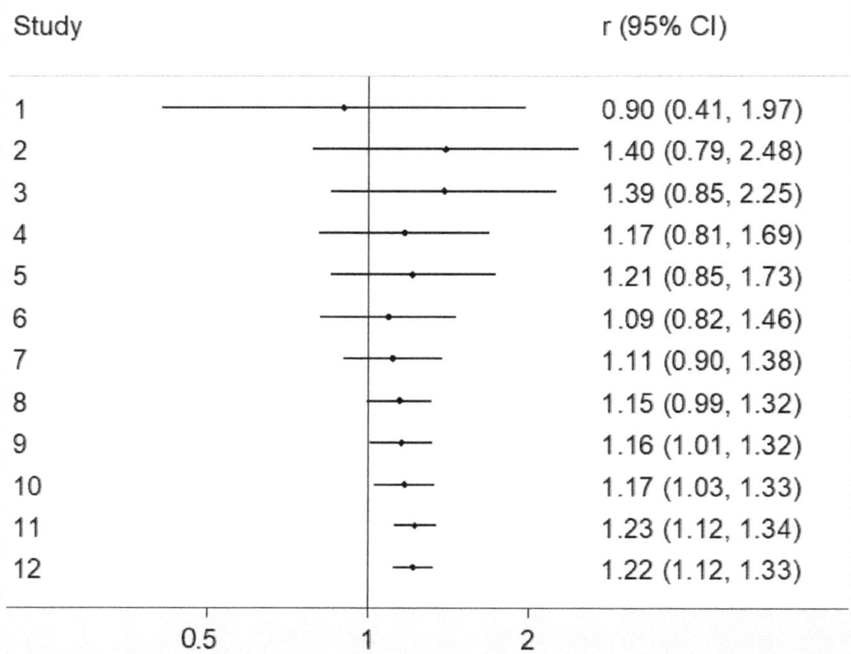

Figure 7. Cumulative analysis for adjusted data.

4. Discussion

In the present meta-analysis, including 12 longitudinal prospective studies and 40,664 older participants, we found that low baseline levels of circulating 25OHD (hypovitaminosis D) were significantly associated with incident diabetes at follow-up (mean of 7.3 years), after adjusting for several potentially relevant confounders.

Noteworthy, none of the prospective studies included reported a significant relationship between baseline hypovitaminosis D and incident diabetes in the multivariable-adjusted analyses. Conversely, when the data were combined in the meta-analysis, including a median of 11 potential confounders, the presence of hypovitaminosis D at baseline was significantly associated with an increased risk of incident diabetes at follow-up by 19%. Our findings suggest that hypovitaminosis D may be a potentially reversible risk factor for the development of diabetes in older adults, who are at the highest risk of both hypovitaminosis D and type 2 diabetes and of their disabling consequences.

As mentioned above, previous studies including adults (aged 50–60 years) did not report unambiguous results regarding the relationship between vitamin D deficiency and type 2 diabetes. Although observational studies showed an inverse association between 25OHD and the risk of incident type 2 diabetes [18], intervention studies were negative or showed limited benefit [19–21]. More recent meta-analyses have reported that vitamin D supplementation significantly reduced the risk of incident type 2 diabetes in patients with prediabetes either considering a moderate to high dose (≥1000 IU/day) [22] or regardless of type, dose, duration, and route of administration [23]. Interestingly, in both of these meta-analyses, the significant association persisted only for nonobese participants and was not significant for obese participants. This is particularly relevant for older adults since, in this population, adiposity appears to be an important factor determining low 25OHD levels [47], which may help explain the results of some studies reporting no association between vitamin D and diabetes in older people included in our meta-analysis. These studies may have been negative or neutral when considered in isolation, but by increasing the statistical power by means of the meta-analysis, the final results showed a significant inverse association between low 25OHD levels and incident diabetes in older adults.

The most recent meta-analysis by Pittas et al. [24] with a very rigorous and well-conducted design showed that supplementation with vitamin D reduced the risk for incident diabetes by 15% in adjusted analyses, reaching a reduction of 76% in participants maintaining 25-OHD levels $\geq$50 ng/mL vs. 20–29 ng/mL during follow-up. These results are outstanding; however, the meta-analysis included a population of adults or not very old people rather than older adults who are at greatest risk of both, hypovitaminosis D [14,25] and type 2 diabetes [1,26].

An observational study based on data from the vitamin D and type 2 diabetes (D2d) study, examined whether intratrial vitamin D exposure affected the risk of incident diabetes in adults with prediabetes. The authors found that the HR for diabetes among participants treated with vitamin D who maintained intratrial 25OHD levels of 100–124 and $\geq$125 nmol/L were 0.48 (0.29–0.80) and 0.29 (0.17–0.50), respectively, compared with those who maintained a level of 50–74 nmol/L, suggesting that vitamin D supplementation should maintain 25OHD serum level $\geq$100 nmol/L ($\geq$40 ng/mL) in order to reduce the risk of diabetes in adults with prediabetes [48].

In a cohort study of 903 adults with a mean age of 74 years, higher 25OHD concentrations (>30 ng/mL) were associated with significantly lower risk for diabetes with an inverse dose-response gradient: each 10 ng/mL higher 25OHD concentration was associated with a 36% lower risk of developing diabetes [49]. This study was not included in the analysis due to a too-long follow-up time (12 years) since the 25OHD measurement.

Another crucial point to be considered in the interpretation of our results is the evidence that low vitamin D levels are associated with a higher risk of mortality [50–52]. Therefore, it is possible that older persons with low serum 25OHD levels may die before evolving into diabetes.

Several mechanisms may help explain our findings of a significant association between low vitamin D levels and an increased risk of incident diabetes, including vitamin D's modulation of insulin secretion [16,53–55]; a reduction of insulin resistance with evidence of vitamin D receptor (VDR) presence in the major insulin-sensitive tissues, i.e., adipocytes [56], muscle [57], and hepatocytes [58]; improved insulin reactivity for glucose transport [59,60]; indirectly through the regulation of calcium [60–64] and magnesium [65–67] metabolism, essential for intracellular processes mediated by insulin; a reduction of chronic inflammation [67–71]; because most patients affected by type 2 diabetes are obese, they may have repression of the bioactivation of vitamin D by cytochrome P450 2R1 [72,73] leading to reduced production of 25OHD; and finally, vitamin D sequestration in excess adipose tissue may blunt vitamin D serum levels [14,74].

Regarding insulin secretion, beta cells express both VDR and 1 alpha-hydroxylase (Cyp27b1), which catalyzes the activation of 25OHD into 1,25$(OH)_2$D, denoting the cell-intrinsic role for VDR [75]. Nevertheless, even if the relationship between vitamin D levels and islet function is significant, whether vitamin D supplementation can directly ameliorate insulin secretion in humans remains uncertain because intervention clinical trials have reported mixed results [75–78]. As regards the ubiquitous expression of VDR, and, in particular, in insulin-sensitive tissues, a study analyzing human and rodent data showed that: (1) 25OHD repletion in insulin-resistant, overweight-to-obese persons was associated with reductions in subcutaneous adipose tissue expression of pro-inflammatory and profibrotic genes, decreased collagen immunofluorescence, and improved hepatic insulin sensitivity as well as worsening trends after six months on the placebo suggesting progressive metabolic effects of 25OHD deficiency. Adipocyte-specific VDR knockout mice following a high-fat diet for 12 weeks mirrored the vitamin D-deficient humans, displaying increased adipose tissue fibrosis and inflammation and hepatic insulin resistance [56].

Because VDR detection in muscle has been controversial, a study examined differences in muscle VDR protein abundance between two mouse strains and between mice and humans. VDR protein was generally lower in some mice models and was higher in humans than in mouse muscle, suggesting that further preclinical and clinical studies are still needed to confirm possible VDR isoforms in human muscle [57]. An in vitro study on hepatocyte ischemia/reperfusion injury showed that VDR mediates hepatocyte apoptosis and proliferation not only through inflammation and immunity, as shown previously, but also through autophagy [58]. As regards glucose transport, it has been reported that 1,25(OH)$_2$D improved glucose uptake via the Sirtuin 1/Insulin receptor substrate 1/glucose transporter 4 (SIRT1/IRS1/(GLUT4) axis by activation of SIRT1, phosphorylation of IRS1, and finally translocation of GLUT4 in myotubes [59]. Moreover, activation of VDR increases calcium concentrations in muscle, enhances the translocation of GLUT4, and increases glucose uptake [60]. Additionally, the pleiotropic role played by the VDR in insulin resistance may comprise: (1) increases in parathyroid hormone (PTH) induced by vitamin D, which increases insulin sensitivity by prompting higher quantities of GLUT1 and GLUT4 in vitamin D-deficient adipose tissue, muscle, and liver [79,80]; (2) suppression of the renin-angiotensin-aldosterone system activity, which damages beta cell function, inhibits peripheral insulin sensitivity [81], delays GLUT4 recruitment [82], and triggers insulin resistance [83].

Vitamin D may regulate insulin secretion and action indirectly through calcium intracellular handling. In the beta cell, 1,25(OH)$_2$D leads to the depolarization of cellular membranes, the opening of calcium channels, and the elevation of intracellular calcium levels [61,64]. A possible molecular mechanism of this effect is due to the fact that 1,25(OH)$_2$D activates protein kinase A and increases channel function by phosphorylating L-type voltage-dependent calcium channel-related proteins [64]. Furthermore, 1,25(OH)$_2$D, through activating VDR, regulates voltage-gated calcium channel expression to improve insulin secretion [84]. An additional mechanism is that 1,25(OH)$_2$D promotes the synthesis of phospholipase C and activates inositol triphosphate that releases calcium from the endoplasmic reticulum (ER) [61,85]. Vitamin D may also regulate calbindin expression [85,86], a calcium-binding protein involved in maintaining calcium concentrations. Regarding magnesium metabolism, the relationship between vitamin D and magnesium is bilateral: several steps in vitamin D metabolism (i.e., 25OHD binding and transport, liver and renal hydroxylation) depend on magnesium as a cofactor [67]. Vitamin D, in turn, plays a key role in magnesium metabolism both by stimulating intestinal magnesium absorption and preventing renal excretion [87]. Thus, it appears that the deficit of each of these two compounds feeds the deficit of the other, leading to a perverse cycle with further worsening of both deficits. The combined deficit of vitamin D and magnesium may lead to clinically relevant outcomes, including not only a higher risk of fragility fractures [88] but also altered insulin secretion and action contributing to an increased risk of incident type 2 diabetes [65,66,89].

Substantial evidence supports the role of vitamin D as an immunomodulatory hormone with notable biological effects on the innate and adaptive immune systems [12]. Vitamin D or overexpression of VDR were reported to suppress cytokine-induced pro-inflammatory responses and apoptosis in beta cell lines and islets [68–70], probably due to the direct suppression of nuclear factor kappa-light-chain-enhancer of activated B cells (NF-kB) activation by VDR. Besides its anti-inflammatory role, vitamin D is also an active suppressor of ER stress and islet amyloid polypeptide-induced beta cell dysfunction [68]. Vitamin D downregulates essential ER stress players in monocytes, the liver, and islets [71].

Finally, the sequestration of vitamin D inside excess adipose tissue decreases 25OHD serum levels [14,74]. There is compelling evidence of the association of an elevated body weight with low circulating 25OHD levels. Vitamin D may impact gene expression by the modulation of inflammation and adipose tissue metabolism in vitro; nevertheless, the exact metabolism of vitamin D in adipose tissue is unknown at present. White adipose tissue expresses VDR and hydroxylase enzymes, but the distribution and concentrations of the generated vitamin D compounds in adipose tissue are largely unidentified. Further options have been proposed in order to measure vitamin D compounds in adipose tissue accurately by using liquid mass spectrometry [74].

Our study has some strengths, such as being one of the few meta-analyses systematically investigating the relationship between vitamin D and incident diabetes in older adults, two relevant aspects of old age. We can also include among the strengths of our study the large number of participants included, the high number of covariates used in the adjusted analyses, a long follow-up period, particularly long considering that older adults have a lower life expectancy, and the low heterogeneity found in almost all of the outcomes included. Nevertheless, the results of the present meta-analysis should be considered taking into account its limitations as well. First, the observational nature of the included studies does not allow us to clarify causative effects and potential pathways responsible for the association of hypovitaminosis D with diabetes. Second, the mean age of the participants included was about 69.1 years. There are currently no studies in the literature on old-old people. Future studies focused on very old people, who are at the highest risk of hypovitaminosis D, are warranted to better understand this association in a vulnerable population. Third, there were a limited number of studies considering each gender; hence, we could not run analyses separated by gender, even if both serum 25OHD and diabetes are generally dissimilar in men and women. Lastly, most of the included studies measured serum 25OHD with RIA, which seems not to be the best method to measure serum 25OHD instead of chemiluminescence [90].

5. Conclusions

The present meta-analysis suggests that hypovitaminosis D is associated with incident diabetes in older adults even after adjusting for several relevant potential confounders, confirming and updating the results of the only previous meta-analysis conducted in older adults in 2017 [27]. This confirms the increasing evidence showing that vitamin D exerts multiple effects beyond bone and calcium metabolism. Since hypovitaminosis D is among the most frequent conditions in old age and the available RCTs investigating the role of vitamin D supplementation in the prevention of type 2 diabetes have been conducted in adults or young older adults, future larger scale well-designed studies are needed to confirm these positive results in old-old populations.

Author Contributions: Conceptualization, L.J.D. and N.V.; methodology, N.V.; data curation, E.M., C.D.P., C.I. and R.R.; formal analysis, N.V.; writing—original draft preparation, L.J.D. and N.V.; writing—review and editing, L.J.D., N.V., E.M., C.D.P., C.I., R.R., C.C., S.G. and M.B. All authors have read and agreed to the published version of the manuscript.

Funding: This research received no external funding.

Conflicts of Interest: The authors declare no conflicts of interest.

References

1. Sun, H.; Saeedi, P.; Karuranga, S.; Pinkepank, M.; Ogurtsova, K.; Duncan, B.B.; Stein, C.; Basit, A.; Chan, J.C.N.; Mbanya, J.C.; et al. IDF Diabetes Atlas: Global, regional and country-level diabetes prevalence estimates for 2021 and projections for 2045. *Diabetes Res. Clin. Pract.* **2022**, *183*, 109119. [CrossRef]
2. Zimmet, P.Z.; Magliano, D.J.; Herman, W.H.; Shaw, J.E. Diabetes: A 21st century challenge. *Lancet Diabetes Endocrinol.* **2014**, *2*, 56–64. [CrossRef]
3. Jacob, L.; Rathmann, W.; Koyanagi, A.; Haro, J.M.; Kostev, K. Association between type 2 diabetes and chronic low back pain in general practices in Germany. *BMJ Open Diabetes Res. Care* **2021**, *9*, e002426. [CrossRef]
4. Ying, A.F.; Tang, T.Y.; Jin, A.; Chong, T.T.; Hausenloy, D.J.; Koh, W.P. Diabetes and other vascular risk factors in association with the risk of lower extremity amputation in chronic limb-threatening ischemia: A prospective cohort study. *Cardiovasc. Diabetol.* **2022**, *21*, 7. [CrossRef]
5. Chaker, L.; Falla, A.; van der Lee, S.J.; Muka, T.; Imo, D.; Jaspers, L.; Colpani, V.; Mendis, S.; Chowdhury, R.; Bramer, W.M.; et al. The global impact of non-communicable diseases on macro-economic productivity: A systematic review. *Eur. J. Epidemiol.* **2015**, *30*, 357–395. [CrossRef]

6. Folkerts, K.; Petruski-Ivleva, N.; Kelly, A.; Fried, L.; Blankenburg, M.; Gay, A.; Kovesdy, C.P. Annual health care resource utilization and cost among type 2 diabetes patients with newly recognized chronic kidney disease within a large U.S. administrative claims database. *J. Manag. Care Spec. Pharm.* **2020**, *26*, 1506–1516. [CrossRef]
7. Park, J.; Zhang, P.; Wang, Y.; Zhou, X.; Look, K.A.; Bigman, E.T. High Out-of-pocket Health Care Cost Burden among Medicare Beneficiaries with Diabetes, 1999–2017. *Diabetes Care* **2021**, *44*, 1797–1804. [CrossRef]
8. Freire, L.B.; Brasil-Neto, J.P.; da Silva, M.L.; Miranda, M.G.C.; de Mattos Cruz, L.; Martins, W.R.; da Silva Paz, L.P. Risk factors for falls in older adults with diabetes mellitus: Systematic review and meta-analysis. *BMC Geriatr.* **2024**, *24*, 201. [CrossRef]
9. Grasset, L.; Frison, E.; Helmer, C.; Catheline, G.; Chene, G.; Dufouil, C. Understanding the relationship between type-2 diabetes MRI markers of neurodegeneration and small vessel disease, and dementia risk: A mediation analysis. *Eur. J. Epidemiol.* **2024**, *39* 409–417. [CrossRef]
10. Titcomb, T.J.; Richey, P.; Casanova, R.; Phillips, L.S.; Liu, S.; Karanth, S.D.; Saquib, N.; Nuno, T.; Manson, J.E.; Shadyab, A.H. et al. Association of type 2 diabetes mellitus with dementia-related and non-dementia-related mortality among postmenopausal women: A secondary competing risks analysis of the women's health initiative. *Alzheimer's Dement.* **2024**, *20*, 234–242. [CrossRef]
11. Reynolds, A.N.; Mann, J. Government inaction and the preventable diabetes pandemic. *Nat. Med.* **2023**, *29*, 791–792. [CrossRef] [PubMed]
12. Charoenngam, N.; Holick, M.F. Immunologic Effects of Vitamin D on Human Health and Disease. *Nutrients* **2020**, *12*, 2097 [CrossRef] [PubMed]
13. Holick, M.F.; Mazzei, L.; Garcia Menendez, S.; Martin Gimenez, V.M.; Al Anouti, F.; Manucha, W. Genomic or Non-Genomic? A Question about the Pleiotropic Roles of Vitamin D in Inflammatory-Based Diseases. *Nutrients* **2023**, *15*, 767. [CrossRef] [PubMed]
14. Dominguez, L.J.; Farruggia, M.; Veronese, N.; Barbagallo, M. Vitamin d sources, metabolism, and deficiency: Available compounds and guidelines for its treatment. *Metabolites* **2021**, *11*, 255. [CrossRef] [PubMed]
15. Taderegew, M.M.; Woldeamanuel, G.G.; Wondie, A.; Getawey, A.; Abegaz, A.N.; Adane, F. Vitamin D deficiency and its associated factors among patients with type 2 diabetes mellitus: A systematic review and meta-analysis. *BMJ Open* **2023**, *13*, e075607 [CrossRef] [PubMed]
16. Lips, P.; Eekhoff, M.; van Schoor, N.; Oosterwerff, M.; de Jongh, R.; Krul-Poel, Y.; Simsek, S. Vitamin D and type 2 diabetes *J. Steroid Biochem. Mol. Biol.* **2017**, *173*, 280–285. [CrossRef]
17. Kawahara, T.; Okada, Y.; Tanaka, Y. Vitamin D efficacy in type 1 and type 2 diabetes. *J. Bone Miner. Metab.* **2024**. [CrossRef] [PubMed]
18. Song, Y.; Wang, L.; Pittas, A.G.; Del Gobbo, L.C.; Zhang, C.; Manson, J.E.; Hu, F.B. Blood 25-hydroxy vitamin D levels and incident type 2 diabetes: A meta-analysis of prospective studies. *Diabetes Care* **2013**, *36*, 1422–1428. [CrossRef] [PubMed]
19. Rosen, C.J.; Adams, J.S.; Bikle, D.D.; Black, D.M.; Demay, M.B.; Manson, J.E.; Murad, M.H.; Kovacs, C.S. The nonskeletal effects of vitamin D: An Endocrine Society scientific statement. *Endocr. Rev.* **2012**, *33*, 456–492. [CrossRef]
20. Sollid, S.T.; Hutchinson, M.Y.; Fuskevag, O.M.; Figenschau, Y.; Joakimsen, R.M.; Schirmer, H.; Njolstad, I.; Svartberg, J. Kamycheva, E.; Jorde, R. No effect of high-dose vitamin D supplementation on glycemic status or cardiovascular risk factors in subjects with prediabetes. *Diabetes Care* **2014**, *37*, 2123–2131. [CrossRef]
21. Pittas, A.G.; Chung, M.; Trikalinos, T.; Mitri, J.; Brendel, M.; Patel, K.; Lichtenstein, A.H.; Lau, J.; Balk, E.M. Systematic review Vitamin D and cardiometabolic outcomes. *Ann. Intern. Med.* **2010**, *152*, 307–314. [CrossRef]
22. Barbarawi, M.; Zayed, Y.; Barbarawi, O.; Bala, A.; Alabdouh, A.; Gakhal, I.; Rizk, F.; Alkasasbeh, M.; Bachuwa, G.; Manson, J.E Effect of Vitamin D Supplementation on the Incidence of Diabetes Mellitus. *J. Clin. Endocrinol. Metab.* **2020**, *105*, 2857–2868 [CrossRef] [PubMed]
23. Zhang, Y.; Tan, H.; Tang, J.; Li, J.; Chong, W.; Hai, Y.; Feng, Y.; Lunsford, L.D.; Xu, P.; Jia, D.; et al. Effects of Vitamin D Supplementation on Prevention of Type 2 Diabetes in Patients with Prediabetes: A Systematic Review and Meta-analysis. *Diabetes Care* **2020**, *43*, 1650–1658. [CrossRef] [PubMed]
24. Pittas, A.G.; Kawahara, T.; Jorde, R.; Dawson-Hughes, B.; Vickery, E.M.; Angellotti, E.; Nelson, J.; Trikalinos, T.A.; Balk, E.M Vitamin D and Risk for Type 2 Diabetes in People with Prediabetes: A Systematic Review and Meta-analysis of Individual Participant Data From 3 Randomized Clinical Trials. *Ann. Intern. Med.* **2023**, *176*, 355–363. [CrossRef] [PubMed]
25. Veronese, N.; Bolzetta, F.; De Rui, M.; Zambon, S.; Corti, M.C.; Musacchio, E.; Baggio, G.; Crepaldi, G.; Perissinotto, E.; Manzato, E.; et al. Serum 25-hydroxyvitamin D and orthostatic hypotension in old people: The Pro.V.A. study. *Hypertension* **2014**, *64*, 481–486. [CrossRef] [PubMed]
26. Sinclair, A.J.; Bellary, S.; Dashora, U.; Abdelhafiz, A.H.; Rowles, S.; Reedman, L.; Turner, B.; Green, M.; Forbes, A.; Middleton, A.; et al. Enhancing diabetes care for the most vulnerable in the 21st century: Interim findings of the National Advisory Panel on Care Home Diabetes (NAPCHD). *Diabet. Med.* **2023**, *40*, e15088. [CrossRef] [PubMed]
27. Lucato, P.; Solmi, M.; Maggi, S.; Bertocco, A.; Bano, G.; Trevisan, C.; Manzato, E.; Sergi, G.; Schofield, P.; Kouidrat, Y.; et al. Low vitamin D levels increase the risk of type 2 diabetes in older adults: A systematic review and meta-analysis. *Maturitas* **2017**, *100*, 8–15. [CrossRef] [PubMed]
28. Von Elm, E.; Altman, D.G.; Egger, M.; Pocock, S.J.; Gøtzsche, P.C.; Vandenbroucke, J.P.; Initiative, S. The Strengthening the Reporting of Observational Studies in Epidemiology (STROBE) Statement: Guidelines for reporting observational studies. *Int. J. Surg.* **2014**, *12*, 1495–1499. [CrossRef] [PubMed]

9. Page, M.J.; McKenzie, J.E.; Bossuyt, P.M.; Boutron, I.; Hoffmann, T.C.; Mulrow, C.D.; Shamseer, L.; Tetzlaff, J.M.; Akl, E.A.; Brennan, S.E. The PRISMA 2020 statement: An updated guideline for reporting systematic reviews. *BMJ* **2021**, *372*, n71. [CrossRef]

0. Luchini, C.; Stubbs, B.; Solmi, M.; Veronese, N. Assessing the quality of studies in meta-analyses: Advantages and limitations of the Newcastle Ottawa Scale. *World J. Meta-Anal.* **2017**, *5*, 80–84. [CrossRef]

1. Higgins, J.P.; Thompson, S.G. Quantifying heterogeneity in a meta-analysis. *Stat. Med.* **2002**, *21*, 1539–1558. [CrossRef]

2. Egger, M.; Smith, G.D.; Schneider, M.; Minder, C. Bias in meta-analysis detected by a simple, graphical test. *BMJ* **1997**, *315*, 629–634. [CrossRef]

3. Duval, S.; Tweedie, R. Trim and fill: A simple funnel-plot–based method of testing and adjusting for publication bias in meta-analysis. *Biometrics* **2000**, *56*, 455–463. [CrossRef]

4. Zhu, A.; Kuznia, S.; Niedermaier, T.; Holleczek, B.; Schöttker, B.; Brenner, H. Consistent inverse associations of total,"bioavailable", free, and "non-bioavailable" vitamin D with incidence of diabetes among older adults with lower baseline HbA1c ($\leq$6%) levels. *Nutrients* **2022**, *14*, 3282. [CrossRef]

5. McCarthy, K.; Laird, E.; O'Halloran, A.M.; Walsh, C.; Healy, M.; Fitzpatrick, A.L.; Walsh, J.B.; Hernández, B.; Fallon, P.; Molloy, A.M. Association between vitamin D deficiency and the risk of prevalent type 2 diabetes and incident prediabetes: A prospective cohort study using data from The Irish Longitudinal Study on Ageing (TILDA). *EClinicalMedicine* **2022**, *53*, 101654. [CrossRef]

6. Xia, J.; Tu, W.; Manson, J.E.; Nan, H.; Shadyab, A.H.; Bea, J.W.; Gower, E.W.; Qi, L.; Cheng, T.-Y.D.; Song, Y. Combined associations of 25-hydroxivitamin D and parathyroid hormone with diabetes risk and associated comorbidities among US white and black women. *Nutr. Diabetes* **2021**, *11*, 29. [CrossRef]

7. Bolland, M.J.; Bacon, C.J.; Horne, A.M.; Mason, B.H.; Ames, R.W.; Wang, T.K.; Grey, A.B.; Gamble, G.D.; Reid, I.R. Vitamin D insufficiency and health outcomes over 5 y in older women. *Am. J. Clin. Nutr.* **2010**, *91*, 82–89. [CrossRef]

8. Kositsawat, J.; Kuchel, G.A.; Tooze, J.A.; Houston, D.K.; Cauley, J.A.; Kritchevsky, S.B.; Strotmeyer, E.S.; Kanaya, A.M.; Harris, T.B.; Johnson, K.C. Vitamin D insufficiency and abnormal hemoglobin A1c in black and white older persons. *J. Gerontol. Ser. A Biomed. Sci. Med. Sci.* **2015**, *70*, 525–531. [CrossRef]

9. Napoli, N.; Schafer, A.L.; Lui, L.-Y.; Cauley, J.A.; Strotmeyer, E.S.; Le Blanc, E.S.; Hoffman, A.R.; Lee, C.G.; Black, D.M.; Schwartz, A.V. Serum 25-hydroxyvitamin D level and incident type 2 diabetes in older men, the Osteoporotic Fractures in Men (MrOS) study. *Bone* **2016**, *90*, 181–184. [CrossRef] [PubMed]

0. Pilz, S.; Van Den Hurk, K.; Nijpels, G.; Stehouwer, C.; Van't Riet, E.; Kienreich, K.; Tomaschitz, A.; Dekker, J. Vitamin D status, incident diabetes and prospective changes in glucose metabolism in older subjects: The Hoorn study. *Nutr. Metab. Cardiovasc. Dis.* **2012**, *22*, 883–889. [CrossRef]

1. Robinson, J.G.; Manson, J.E.; Larson, J.; Liu, S.; Song, Y.; Howard, B.V.; Phillips, L.; Shikany, J.M.; Allison, M.; Curb, J.D. Lack of association between 25 (OH) D levels and incident type 2 diabetes in older women. *Diabetes Care* **2011**, *34*, 628–634. [CrossRef] [PubMed]

2. Schafer, A.; Napoli, N.; Lui, L.; Schwartz, A.; Black, D.; Study of Osteoporotic Fractures. Serum 25-hydroxyvitamin D concentration does not independently predict incident diabetes in older women. *Diabet. Med.* **2014**, *31*, 564–569. [CrossRef] [PubMed]

3. Schöttker, B.; Herder, C.; Rothenbacher, D.; Perna, L.; Müller, H.; Brenner, H. Serum 25-hydroxyvitamin D levels and incident diabetes mellitus type 2: A competing risk analysis in a large population-based cohort of older adults. *Eur. J. Epidemiol.* **2013**, *28*, 267–275. [CrossRef] [PubMed]

4. Thorand, B.; Zierer, A.; Huth, C.; Linseisen, J.; Meisinger, C.; Roden, M.; Peters, A.; Koenig, W.; Herder, C. Effect of serum 25-hydroxyvitamin D on risk for type 2 diabetes may be partially mediated by subclinical inflammation: Results from the MONICA/KORA Augsburg study. *Diabetes Care* **2011**, *34*, 2320–2322. [CrossRef] [PubMed]

5. Veronese, N.; Sergi, G.; De Rui, M.; Bolzetta, F.; Toffanello, E.D.; Zambon, S.; Corti, M.-C.; Sartori, L.; Musacchio, E.; Baggio, G. Serum 25-hydroxyvitamin D and incidence of diabetes in elderly people: The PRO. VA study. *J. Clin. Endocrinol. Metab.* **2014**, *99*, 2351–2358. [CrossRef] [PubMed]

6. Jorde, R.; Sneve, M.; Hutchinson, M.; Emaus, N.; Figenschau, Y.; Grimnes, G. Tracking of serum 25-hydroxyvitamin D levels during 14 years in a population-based study and during 12 months in an intervention study. *Am. J. Epidemiol.* **2010**, *171*, 903–908. [CrossRef] [PubMed]

7. Ceolin, G.; Matsuo, L.H.; Ocker, G.; Grellert, M.; d'Orsi, E.; Venske, D.K.R.; Moreira, J.D. Adiposity and physical activity are among the main determinants of serum vitamin D concentrations in older adults: The EpiFloripa Aging Cohort Study. *Nutr. Res.* **2023**, *111*, 59–72. [CrossRef] [PubMed]

8. Dawson-Hughes, B.; Staten, M.A.; Knowler, W.C.; Nelson, J.; Vickery, E.M.; LeBlanc, E.S.; Neff, L.M.; Park, J.; Pittas, A.G. Intratrial exposure to vitamin D and new-onset diabetes among adults with prediabetes: A secondary analysis from the vitamin D and type 2 diabetes (D2d) study. *Diabetes Care* **2020**, *43*, 2916–2922. [CrossRef]

9. Park, S.K.; Garland, C.F.; Gorham, E.D.; BuDoff, L.; Barrett-Connor, E. Plasma 25-hydroxyvitamin D concentration and risk of type 2 diabetes and pre-diabetes: 12-year cohort study. *PLoS ONE* **2018**, *13*, e0193070. [CrossRef]

0. Schottker, B.; Ball, D.; Gellert, C.; Brenner, H. Serum 25-hydroxyvitamin D levels and overall mortality. A systematic review and meta-analysis of prospective cohort studies. *Ageing Res. Rev.* **2013**, *12*, 708–718. [CrossRef]

1. Samefors, M.; Ostgren, C.J.; Molstad, S.; Lannering, C.; Midlov, P.; Tengblad, A. Vitamin D deficiency in elderly people in Swedish nursing homes is associated with increased mortality. *Eur. J. Endocrinol.* **2014**, *170*, 667–675. [CrossRef] [PubMed]

52. Rejnmark, L.; Bislev, L.S.; Cashman, K.D.; Eiriksdottir, G.; Gaksch, M.; Grubler, M.; Grimnes, G.; Gudnason, V.; Lips, P.; Pilz, S et al. Non-skeletal health effects of vitamin D supplementation: A systematic review on findings from meta-analyses summarizin trial data. *PLoS ONE* **2017**, *12*, e0180512. [CrossRef] [PubMed]
53. Norman, A.W.; Frankel, J.B.; Heldt, A.M.; Grodsky, G.M. Vitamin D deficiency inhibits pancreatic secretion of insulin. *Scienc* **1980**, *209*, 823–825. [CrossRef] [PubMed]
54. Cade, C.; Norman, A.W. Vitamin D3 improves impaired glucose tolerance and insulin secretion in the vitamin D-deficient ra in vivo. *Endocrinology* **1986**, *119*, 84–90. [CrossRef] [PubMed]
55. Lemieux, P.; Weisnagel, S.J.; Caron, A.Z.; Julien, A.S.; Morisset, A.S.; Carreau, A.M.; Poirier, J.; Tchernof, A.; Robitaille, J.; Bergero J.; et al. Effects of 6-month vitamin D supplementation on insulin sensitivity and secretion: A randomised, placebo-controlle trial. *Eur. J. Endocrinol.* **2019**, *181*, 287–299. [CrossRef] [PubMed]
56. Lontchi-Yimagou, E.; Kang, S.; Goyal, A.; Zhang, K.; You, J.Y.; Carey, M.; Jain, S.; Bhansali, S.; Kehlenbrink, S.; Guo, P.; et a Insulin-sensitizing effects of vitamin D repletion mediated by adipocyte vitamin D receptor: Studies in humans and mice. *Mc Metab.* **2020**, *42*, 101095. [CrossRef] [PubMed]
57. Lalunio, H.; Parker, L.; Hanson, E.D.; Gregorevic, P.; Levinger, I.; Hayes, A.; Goodman, C.A. Detecting the vitamin D recepto (VDR) protein in mouse and human skeletal muscle: Strain-specific, species-specific and inter-individual variation. *Mol. Ce Endocrinol.* **2023**, *578*, 112050. [CrossRef] [PubMed]
58. Fang, M.; Zhong, C. Vitamin D Receptor Regulates Autophagy to Inhibit Apoptosis and Promote Proliferation in Hepatocyt Injury. *J. Nippon. Med. Sch.* **2023**, *90*, 89–95. [CrossRef]
59. Manna, P.; Achari, A.E.; Jain, S.K. Vitamin D supplementation inhibits oxidative stress and upregulate SIRT1/AMPK/GLUT cascade in high glucose-treated 3T3L1 adipocytes and in adipose tissue of high fat diet-fed diabetic mice. *Arch. Biochem. Biophy* **2017**, *615*, 22–34. [CrossRef]
60. Wright, D.C.; Hucker, K.A.; Holloszy, J.O.; Han, D.H. Ca^{2+} and AMPK both mediate stimulation of glucose transport by muscl contractions. *Diabetes* **2004**, *53*, 330–335. [CrossRef]
61. Altieri, B.; Grant, W.B.; Della Casa, S.; Orio, F.; Pontecorvi, A.; Colao, A.; Sarno, G.; Muscogiuri, G. Vitamin D and pancrea The role of sunshine vitamin in the pathogenesis of diabetes mellitus and pancreatic cancer. *Crit. Rev. Food Sci. Nutr.* **2017**, *5* 3472–3488. [CrossRef] [PubMed]
62. Sergeev, I.N.; Rhoten, W.B. 1,25-Dihydroxyvitamin D3 evokes oscillations of intracellular calcium in a pancreatic beta-cell lin *Endocrinology* **1995**, *136*, 2852–2861. [CrossRef] [PubMed]
63. Doyle, M.E.; Egan, J.M. Pharmacological agents that directly modulate insulin secretion. *Pharmacol. Rev.* **2003**, *55*, 105–13 [CrossRef] [PubMed]
64. Gilon, P.; Chae, H.Y.; Rutter, G.A.; Ravier, M.A. Calcium signaling in pancreatic beta-cells in health and in Type 2 diabetes. *Ce Calcium* **2014**, *56*, 340–361. [CrossRef] [PubMed]
65. Barbagallo, M.; Veronese, N.; Dominguez, L.J. Magnesium in Aging, Health and Diseases. *Nutrients* **2021**, *13*, 463. [CrossRe [PubMed]
66. Dominguez, L.J.; Veronese, N.; Barbagallo, M. Magnesium and the Hallmarks of Aging. *Nutrients* **2024**, *16*, 496. [CrossRe [PubMed]
67. Dominguez, L.J.; Veronese, N.; Guerrero-Romero, F.; Barbagallo, M. Magnesium in Infectious Diseases in Older People. *Nutrien* **2021**, *13*, 180. [CrossRef] [PubMed]
68. Wei, Z.; Yoshihara, E.; He, N.; Hah, N.; Fan, W.; Pinto, A.F.M.; Huddy, T.; Wang, Y.; Ross, B.; Estepa, G.; et al. Vitamin D Switche BAF Complexes to Protect beta Cells. *Cell* **2018**, *173*, 1135–1149.e15. [CrossRef] [PubMed]
69. Chen, C.; Luo, Y.; Su, Y.; Teng, L. The vitamin D receptor (VDR) protects pancreatic beta cells against Forkhead box class O (FOXO1)-induced mitochondrial dysfunction and cell apoptosis. *Biomed. Pharmacother.* **2019**, *117*, 109170. [CrossRef]
70. Morro, M.; Vila, L.; Franckhauser, S.; Mallol, C.; Elias, G.; Ferre, T.; Molas, M.; Casana, E.; Rodo, J.; Pujol, A.; et al. Vitamin I Receptor Overexpression in beta-Cells Ameliorates Diabetes in Mice. *Diabetes* **2020**, *69*, 927–939. [CrossRef]
71. Riek, A.E.; Oh, J.; Sprague, J.E.; Timpson, A.; de las Fuentes, L.; Bernal-Mizrachi, L.; Schechtman, K.B.; Bernal-Mizrachi, C Vitamin D suppression of endoplasmic reticulum stress promotes an antiatherogenic monocyte/macrophage phenotype in type diabetic patients. *J. Biol. Chem.* **2012**, *287*, 38482–38494. [CrossRef] [PubMed]
72. Roizen, J.D.; Long, C.; Casella, A.; O'Lear, L.; Caplan, I.; Lai, M.; Sasson, I.; Singh, R.; Makowski, A.J.; Simmons, R.; et al. Obesit Decreases Hepatic 25-Hydroxylase Activity Causing Low Serum 25-Hydroxyvitamin D. *J. Bone Miner. Res.* **2019**, *34*, 1068–107 [CrossRef] [PubMed]
73. Elkhwanky, M.S.; Kummu, O.; Piltonen, T.T.; Laru, J.; Morin-Papunen, L.; Mutikainen, M.; Tavi, P.; Hakkola, J. Obesity Represse CYP2R1, the Vitamin D 25-Hydroxylase, in the Liver and Extrahepatic Tissues. *JBMR Plus* **2020**, *4*, e10397. [CrossRef] [PubMed
74. Spyksma, E.E.; Alexandridou, A.; Mai, K.; Volmer, D.A.; Stokes, C.S. An Overview of Different Vitamin D Compounds in th Setting of Adiposity. *Nutrients* **2024**, *16*, 231. [CrossRef] [PubMed]
75. Inomata, S.; Kadowaki, S.; Yamatani, T.; Fukase, M.; Fujita, T. Effect of 1 alpha (OH)-vitamin D3 on insulin secretion in diabete mellitus. *Bone Miner.* **1986**, *1*, 187–192. [PubMed]
76. Borissova, A.M.; Tankova, T.; Kirilov, G.; Dakovska, L.; Kovacheva, R. The effect of vitamin D3 on insulin secretion and periphera insulin sensitivity in type 2 diabetic patients. *Int. J. Clin. Pr.* **2003**, *57*, 258–261. [CrossRef]

77. Nyomba, B.L.; Auwerx, J.; Bormans, V.; Peeters, T.L.; Pelemans, W.; Reynaert, J.; Bouillon, R.; Vantrappen, G.; De Moor, P. Pancreatic secretion in man with subclinical vitamin D deficiency. *Diabetologia* **1986**, *29*, 34–38. [CrossRef] [PubMed]
78. Rasouli, N.; Brodsky, I.G.; Chatterjee, R.; Kim, S.H.; Pratley, R.E.; Staten, M.A.; Pittas, A.G.; Group, D.d.R. Effects of Vitamin D Supplementation on Insulin Sensitivity and Secretion in Prediabetes. *J. Clin. Endocrinol. Metab.* **2022**, *107*, 230–240. [CrossRef] [PubMed]
79. Ni, Z.; Smogorzewski, M.; Massry, S.G. Effects of parathyroid hormone on cytosolic calcium of rat adipocytes. *Endocrinology* **1994**, *135*, 1837–1844. [CrossRef] [PubMed]
80. Thomas, D.M.; Rogers, S.D.; Sleeman, M.W.; Pasquini, G.M.; Bringhurst, F.R.; Ng, K.W.; Zajac, J.D.; Best, J.D. Modulation of glucose transport by parathyroid hormone and insulin in UMR 106-01, a clonal rat osteogenic sarcoma cell line. *J. Mol. Endocrinol.* **1995**, *14*, 263–275. [CrossRef]
81. Cheng, Q.; Boucher, B.J.; Leung, P.S. Modulation of hypovitaminosis D-induced islet dysfunction and insulin resistance through direct suppression of the pancreatic islet renin-angiotensin system in mice. *Diabetologia* **2013**, *56*, 553–562. [CrossRef]
82. Muscogiuri, G.; Chavez, A.O.; Gastaldelli, A.; Perego, L.; Tripathy, D.; Saad, M.J.; Velloso, L.; Folli, F. The crosstalk between insulin and renin-angiotensin-aldosterone signaling systems and its effect on glucose metabolism and diabetes prevention. *Curr. Vasc. Pharmacol.* **2008**, *6*, 301–312. [CrossRef]
83. Wei, Y.; Sowers, J.R.; Clark, S.E.; Li, W.; Ferrario, C.M.; Stump, C.S. Angiotensin II-induced skeletal muscle insulin resistance mediated by NF-kappaB activation via NADPH oxidase. *Am. J. Physiol. Endocrinol. Metab.* **2008**, *294*, E345–E351. [CrossRef] [PubMed]
84. Kjalarsdottir, L.; Tersey, S.A.; Vishwanath, M.; Chuang, J.C.; Posner, B.A.; Mirmira, R.G.; Repa, J.J. 1,25-Dihydroxyvitamin D(3) enhances glucose-stimulated insulin secretion in mouse and human islets: A role for transcriptional regulation of voltage-gated calcium channels by the vitamin D receptor. *J. Steroid Biochem. Mol. Biol.* **2019**, *185*, 17–26. [CrossRef]
85. De Boland, A.R.; Boland, R.L. Non-genomic signal transduction pathway of vitamin D in muscle. *Cell. Signal.* **1994**, *6*, 717–724. [CrossRef] [PubMed]
86. Morrissey, R.L.; Bucci, T.J.; Richard, B.; Empson, N.; Lufkin, E.G. Calcium-binding protein: Its cellular localization in jejunum, kidney and pancreas. *Proc. Soc. Exp. Biol. Med.* **1975**, *149*, 56–60. [CrossRef]
87. Hardwick, L.L.; Jones, M.R.; Brautbar, N.; Lee, D.B. Magnesium absorption: Mechanisms and the influence of vitamin D, calcium and phosphate. *J. Nutr.* **1991**, *121*, 13–23. [CrossRef]
88. Veronese, N.; Stubbs, B.; Solmi, M.; Noale, M.; Vaona, A.; Demurtas, J.; Maggi, S. Dietary magnesium intake and fracture risk: Data from a large prospective study. *Br. J. Nutr.* **2017**, *117*, 1570–1576. [CrossRef] [PubMed]
89. Veronese, N.; Dominguez, L.J.; Pizzol, D.; Demurtas, J.; Smith, L.; Barbagallo, M. Oral magnesium supplementation for treating glucose metabolism parameters in people with or at risk of diabetes: A systematic review and meta-analysis of double-blind randomized controlled trials. *Nutrients* **2021**, *13*, 4074. [CrossRef] [PubMed]
90. Holick, M.F. Vitamin D status: Measurement, interpretation, and clinical application. *Ann. Epidemiol.* **2009**, *19*, 73–78. [CrossRef]

Disclaimer/Publisher's Note: The statements, opinions and data contained in all publications are solely those of the individual author(s) and contributor(s) and not of MDPI and/or the editor(s). MDPI and/or the editor(s) disclaim responsibility for any injury to people or property resulting from any ideas, methods, instructions or products referred to in the content.

nutrients

MDPI

Article

Effect of 2 Years of Monthly Calcifediol Administration in Postmenopausal Women with Vitamin D Insufficiency

Marco Occhiuto [1,†], Jessica Pepe [1,†], Luciano Colangelo [1], Marco Lucarelli [2], Antonio Angeloni [2], Luciano Nieddu [3], Viviana De Martino [1], Salvatore Minisola [1,*] and Cristiana Cipriani [1]

1 Department of Clinical, Internal, Anaesthesiologic and Cardiovascular Sciences, Sapienza University of Rome, Viale del Policlinico 155, 00161 Rome, Italy; marco.occhiuto@uniroma1.it (M.O.); jessica.pepe@uniroma1.it (J.P.); luciano.colangelo@uniroma1.it (L.C.); viviana.demartino@uniroma1.it (V.D.M.); cristiana.cipriani@gmail.com (C.C.)

2 Department of Experimental Medicine, Sapienza University of Rome, Viale del Policlinico 155, 00161 Rome, Italy; marco.lucarelli@uniroma1.it (M.L.); antonio.angeloni@uniroma1.it (A.A.)

3 Department of Humanistic and Social International Sciences, UNINT University, Via Cristoforo Colombo 200, 00147 Rome, Italy; l.nieddu@gmail.com

* Correspondence: salvatore.minisola@uniroma1.it

† These authors contributed equally to the work.

Abstract: Background: We assessed the long-term (24 months) efficacy and safety of monthly calcifediol (0.266 mg) in the correction and maintenance of total 25(OH)D levels in postmenopausal women with basal values <30 ng/mL. Methods: We initially enrolled 45 consecutive patients during the period September 2019–September 2020. After an initial visit, patients were instructed to return at 3, 6, 9, 12 and 24 months for measuring serum total 25(OH)D, ionised calcium, creatinine and isoenzyme of alkaline phosphatase (bALP). Here, we report only the per-protocol analysis, because the COVID-19 pandemic precluded adherence to the scheduled visits for some patients. Results: The patients' mean age was 62.4 ± 9.0 years. Mean basal 25(OH)D levels were 20.5 ± 5.3 ng/mL. There was a continuous increase of mean 25(OH)D values (p for trend < 0.001). However, mean values at month 24 (36.7 ± 15.9) were not significantly different in respect to values at month 12 (41.2 ± 11.18). At 24 months, only 1 out 19 patients had a value <20 ng/mL. There was a significant decrease with time of mean values of bALP ($p < 0.0216$), with no significant changes between 12 and 24 months. No significant changes were observed as far as ionised calcium or creatinine were concerned. Conclusions: The long-term administration of calcifediol maintains stable and sustained 25(OH)D concentrations, with no safety concerns.

Keywords: calcifediol; vitamin D; ionised calcium; bone alkaline phosphatase

1. Introduction

Vitamin D insufficiency (and deficiency) is largely prevalent in the world [1]. If we consider a threshold of 12 ng/mL (30 nmol/L; to convert ng/mL to nmol/L multiply the ng/mL by 2.5), it has been estimated that about 5% of the American population and about 9% of the Canadian population have serum values of 25(OH)D below this level [2]. Furthermore, the risk of being vitamin D-deficient in the United States is higher in the black population. Based on the same thresholds, about 13% of the population is at risk of vitamin D deficiency in Europe. This percentage can increase up to 28–65% if we take into consideration ethnic minority groups within Northern European countries [2]. However, the prevalence of serum 25(OH)D below 20 ng/mL raises in the range of 27.2 to 61.4%. Considering both continents, it has been estimated that 120 million subjects are vitamin D-deficient using a 12 ng/mL threshold, while there are 390 million subjects with 25(OH)D below 20 nmol/L. These figures are astonishing and are a call to action for specific initiatives to eliminate this "global pandemic". However, even though many such initiatives have been undertaken over the last few years to increase awareness about this clinical problem, the situation seems to be unchanged.

Low values of 25(OH)D, considered the best biological marker of body stores of vitamin D from all sources, have detrimental effects on the skeleton and have been also associated with a number of non-skeletal outcomes [3–5]. Indeed, a huge number of epidemiological studies have consistently demonstrated that low 25(OH)D values are associated with an increased risk of cancer and chronic diseases, among which are cardiovascular, neurological, autoimmune and infectious diseases. Some studies have also found an association with an increased risk of mortality [6].

Vitamin D is synthesised from 7-dehydrocholesterol in the skin by absorbing ultraviolet B (UVB) radiation. Vitamin D can also be obtained from the diet or by means of dietary supplementation in the form of either vitamin D3 (cholecalciferol) or D2 (ergocalciferol). Within a few hours of being synthesised in the skin or taken with foods, it undergoes two consecutive hydroxylation steps: the first one is represented by the addition of one hydroxyl group in position 25 to generate 25(OH)D in the liver; the second step involves the addition of another hydroxyl group in position 1 in the kidney to generate the final metabolite, that is, $1,25(OH)_2D$. The most important function of the final metabolite is an effect on intestinal calcium absorption. Without vitamin D, only 10–15% of dietary calcium and about 60% of phosphorus are absorbed. Vitamin D sufficiency enhances calcium and phosphorus absorption by 30–40% and 80%, respectively. Other tissues contain the enzymes needed for hydroxylation, but the liver is the main source for 25-hydroxylation, and the kidney is the main source for 1α-hydroxylation. Control of the metabolism of vitamin D to its active metabolite, $1,25(OH)_2D$, is exerted primarily at the renal level, where calcium, phosphorus, parathyroid hormone, fibroblast growth factor 23 (FGF23) and $1,25(OH)_2D$ regulate the levels of $1,25(OH)_2D$ that are produced.

There are essentially four possibilities for increasing circulating levels of 25(OH)D, each carrying inherent advantages and disadvantages. Firstly, by increasing skin synthesis of vitamin D with exposure to UVB light. Secondly, by increasing the intake of vitamin D foods; however, few foods contain vitamin D. The flesh of fatty fish (that is, trout, salmon, tuna and mackerel) and fish oils are among the best sources. Beef liver, egg yolks and cheese have small amounts of vitamin D, primarily in the form of vitamin D3 and its metabolite 25(OH)D3. Mushrooms provide variable amounts of vitamin D2. Thirdly, by enriching food with vitamin D or 25(OH)D and finally, by pharmacological supplementation [7]. Concerning the last option, there are a number of commercial products that can be utilised [8]; they basically recapitulate the biosynthetic pathway of vitamin D from its origin to the final metabolite. The most utilised of these products are ergocalciferol, cholecalciferol, calcifediol, 1α(OH)D and calcitriol, even though their utilisation is markedly different in European and American countries. Other compounds are utilised in specific clinical conditions, such as in chronic kidney disease [9].

Calcifediol is the intermediate metabolite in the metabolic pathway that leads from cholecalciferol to calcitriol production, being formed by the addition of a 25-hydroxyl group to its precursor (mainly by the enzymes microsomal CYP2R1 and mitochondrial CYP27A1). Calcifediol binds to Vitamin D-binding protein in the circulation. It has distinct metabolic and pharmacokinetics profiles, compared to both its precursor and downstream final metabolite [10]. A number of previous investigations have shown that calcifediol administration determines a rapid increase of circulating 25(OH)D compared to cholecalciferol [11]. Indeed, calcifediol is rapidly absorbed by intestinal cells and transported through the portal vein, and therefore, it is immediately available in the circulation. Native vitamin D is transported more slowly by chylomicrons via the lymphatic system. The rapid absorption and its independence from hepatic 25-hydroxylation lead to pharmacokinetics that make calcifediol an immediately available substrate for the synthesis of calcitriol, either for systemic transport or for local paracrine or autocrine actions. In this context, Charoenngam et al. [11] evaluated the effect of cholecalciferol and calcifediol in 6 patients with malabsorption and 10 healthy subjects, reporting thought-provoking pharmacokinetic data.

After administration of 900 mcg of cholecalciferol, the area under the curve was 68% lower for patients with malabsorption than for healthy patients. However, the administration of 900 mcg of calcifediol revealed no differences between the two groups. This suggests that the bioavailability of 900 mcg of orally administered 25(OH)D_3 was not different between malabsorptive patients and healthy participants.

Concerning long-term effects, the longest time frame of calcifediol investigation has been 12 months, with studies differing in both the amount and schedule of administration [12–14]. In the longest-lasting study to date, the monthly administration of 0.266 mg of calcifediol for as long as one year produced stable and sustained 25(OH)D concentrations with no associated safety concerns [12].

This study has been therefore carried out to investigate the 2-year effect of monthly 0.266 mg calcifediol administration in restoring and maintaining 25(OH)D levels in postmenopausal females with hypovitaminosis D.

2. Materials and Methods

This is an open-label study in which patients received calcifediol at a dose of 0.266 mg monthly (Neodidro soft gelatine capsule, Bruno Farmaceutici S. p. A., Rome, Italy) for a period of two years. In addition to being a response to the lack of long-term data (i.e., more than one year), this schedule was chosen because it is the one suggested by the Italian Medical Agency (note 96 of the Agenzia Italiana del Farmaco, AIFA, Rome, Italy) for vitamin D replenishment when circulating values are between 13 and 20 ng/mL or above 20 ng/mL. Furthermore, this type of administration aims, on the one hand, at avoiding any serious adverse events while, on the other hand, guaranteeing a good adherence in respect to daily or weekly administration. All participants signed their informed consent prior to entering the investigation.

We initially enrolled 45 consecutive postmenopausal patients with total 25(OH)D values less than 30 ng/mL, during the period September 2019–September 2020. Being postmenopausal was defined as amenorrhea for more than 6 months or follicle-stimulating hormone levels above 30 I. U./L, with oestradiol values less than 30 pg/mL) They attended our mineral metabolism unit because they had already carried out a bone mineral density (BMD) measurement or because they were worried about a possible diagnosis of osteoporosis. Exclusion criteria were medical or therapeutic conditions that could interfere with our investigation. These were represented by the following: progressive major illnesses, severe malabsorption syndrome, stage IV chronic kidney disease (that is, an estimated glomerular filtration rate between 15 and 29 mL/min, together with moderate to severe kidney damage), nephrolithiasis, Paget's disease of the bone, primary hyperparathyroidism, genetic and acquired hypoparathyroidism, hypercalcaemia, hypothyroidism and hyperthyroidism, sarcoidosis, hypercalciuria and intolerance to calcifediol. Patients taking treatments interfering with bone and mineral metabolism such as glucocorticoids, diuretics, lithium, immunosuppressants, antiretroviral therapy and other drugs possibly interfering with vitamin D absorption and catabolism were also excluded. The initial visit included a physical examination, blood collection and BMD imaging of the spine and hip with a vertebral fracture assessment, as previously described [15]. An experienced skeletal radiologist evaluated the presence of fractures according to an algorithm-based qualitative approach. Both the vertebral shape and the appearance of the endplate were evaluated, to clearly differentiate vertebral fractures from developmental variants, degenerative change and so on. Patients were therefore instructed to return at 3, 6, 9, 12 and 24 months for measuring serum total 25(OH)D, ionised calcium, creatinine and isoenzymes of alkaline phosphatase (bALP). Professional nurses obtained blood samples at each visit and carried out a medical history for possible intercurrent illnesses, changes in the usual therapy and possible side effects.

2.1. Biochemical Analysis

Briefly, blood samples were taken between 8.00 and 10.00 a.m. for the measurement of serum ionised calcium (Ca^{++}), 25-hydroxyvitamin D, creatinine and isoenzymes of alkaline phosphatase in the fasting state. Details for the measurement of blood chemistries and bone mineral density have been previously described [16,17]. More specifically, except for Ca^{++}, the level of which was measured immediately after collection (within 2 h of sampling, while the room temperature was kept at 4 °C), aliquoted blood samples were stored at −80 °C and assayed at a later time in batches. Serum Ca^{++} determination was carried out using an ion-selective electrode with the fully automated biochemical analyser NOVA 8 (Nova Biomedical, Waltham, MA, USA); 25(OH)D was measured by chemiluminescence–immunoassay (CLIA) with the fully automated LIAISON® analyser (Diasorin S.p.A. Saluggia, Italy). Due to technical limitations imposed by use of the LIAISON® instrument, the 25(OH)D assays were performed on a quarterly basis. The determinations of serum 25(OH)D were performed with a competitive one-step backfill chemiluminescence assay (Vitamin D TOTAL Assay, DiaSorin USA, Stillwater, MN, USA) having a measurement range of 4–150 ng/mL and functional sensitivity ≤4.0 ng/mL; the intra- and inter-assay precision were 8.9% and 12.8%, respectively, with a reported 100% detection of both 25-hydroxyvitamin D2 and 25-hydroxyvitamin D3. Serum creatinine was measured by a Cobas c501 analyser (Cobas® Roche Diagnostics S.p.A., Monza, Italy). The bone isoenzyme of alkaline phosphatase was measured by IDS-iSYS Ostase® BAP assay (Immunodiagnostics Systems, Boldon, UK).

Safety and tolerability were also examined throughout the study. This study was performed in accordance with the Declaration of Helsinki, as well as local laws and regulations. The ethical committee of our university reviewed and approved the protocol.

2.2. Statistical Analysis

Quantitative variables are summarised by mean ± standard deviation. Student's t test or the Mann–Whitney test was used for pairwise comparisons as appropriate, depending on the characteristics of the data. The effect of time on the trend of each single outcome has been assessed using linear mixed-effect models for repeated measures. Post hoc pairwise comparisons were performed to assess the significance of the differences of the outcomes at various times. For statistical significance, a p value < 0.05 was considered appropriate.

3. Results

We initially enrolled 45 consecutive postmenopausal patients attending our mineral metabolism service who satisfied our inclusion and exclusion criteria, with serum 25(OH)D levels below 30 ng/mL. Thirty-six of them initially agreed to participate. The enrolment period (September 2019–September 2020) and the follow-up observation period coincided with the beginning of the COVID-19 pandemic in Italy and subsequent restrictions on mobility. In respect to the initial protocol, this had two important consequences: first, not all the enrolled patients ended the protocol, and secondly, not all patients adhered to the scheduled visits, mainly because they were afraid of coming to our hospital, which was a referral centre for COVID-19. We therefore report the results according to a per-protocol analysis.

Table 1 shows the anthropometric, biochemical and densitometric parameters of the 36 patients who initially adhered to the study protocol. There were no significant statistical differences in respect to those that declined to participate.

The mean (±1 SD) age of the patients was 62.4 ± 9 years and about 18% of them had one or more vertebral fractures. Basal total 25(OH)D levels were 20.9 ± 4.2 ng/mL. Two patients were already taking generic alendronate (without vitamin D), and a further nine patients started taking various therapies for osteoporosis during the study, owing to fractures or a high risk of fractures as determined by FRAX®. The mean T-score values at the three sites considered (that is, lumbar spine. femoral neck and total hip) were in the osteopenic range. Twelve patients were treated for essential hypertension, two for type

2 diabetes mellitus and nine for hyperlipidaemia. None of them was taking drugs that could interfere with vitamin D metabolism.

Table 1. Main anthropometric, biochemical and densitometric parameters of patients investigated.

Age (years)	62.4 (±9)
Weight (kg)	60.7 (±9.5)
Height (cm)	161.8 (±7.1)
Body Mass Index (kg/m^2)	23.1 (±3.4)
Years since menopause	14 (±8.7)
Vertebral fractures	17.6%
L1–L4 BMD (g/cm^2)	0.840 (±0.161)
L1–L4 T-score (SD)	−2.0 (±1.3)
Femoral neck BMD (g/cm^2)	0.654 (±0.102)
Femoral neck T-score (SD)	−1.92 (±0.86)
Femoral total BMD (g/cm^2)	0.765 (±0.124)
Femoral total T-score (SD)	−1.48 (±1.05)
25 OH vitamina D (ng/mL)	20.5 (±5.3)
Ionised calcium (mmol/L)	1.26 (±0.03)
Total calcium (mg/dL)	9.2 (±0.7)
Phosphate (mg/dL)	3.7 (±0.3)
Creatinine (mg/dL)	0.74 (±0.17)
Bone alkaline phosphatase (μg/L)	22.9 (±10.4)

There was a continuous increase in mean 25(OH)D values (p for trend < 0.001) during the 2-year observation study. However, the mean values at month 24 (36.7 ± 15.9) were not significantly different in respect to values at month 12 (41.2 ± 11.18) (Figure 1). At the end of the study (i.e., 24 months) only 1 out 19 patients had a value less than 20 ng/mL, and 5 patients had values above 20 ng mL.

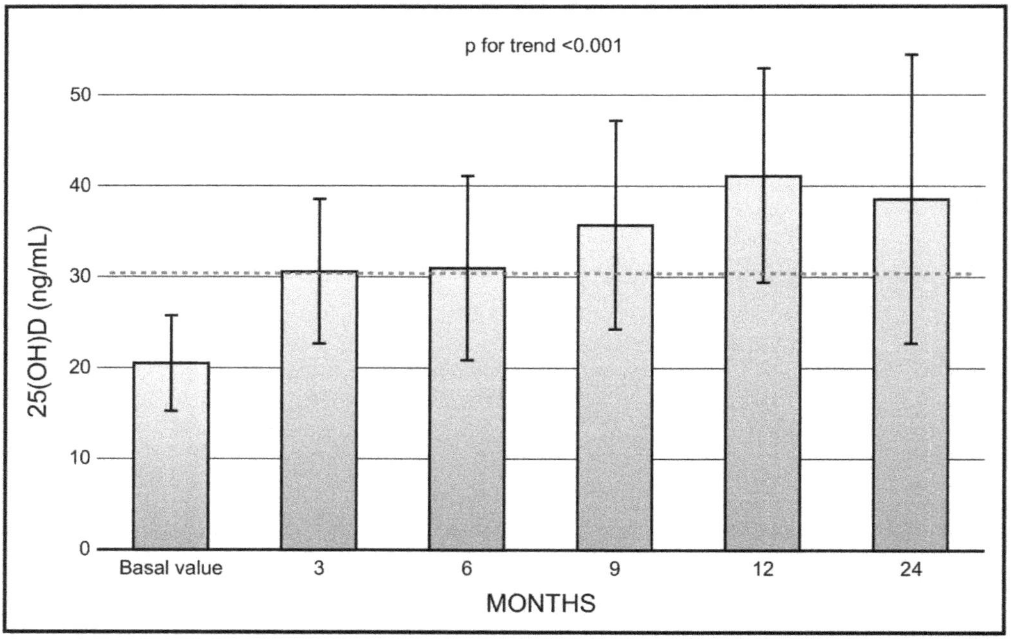

Figure 1. Time course of serum 25(OH)D following monthly administration of 0.266 mg calcifediol. Error bars indicate 1 SD.

There was a significant decrease with time in mean values of bone ALP ($p < 0.0216$), with no significant changes in mean values between 12 and 24 months (22.9 ± 11.9 vs. 20.8 ± 20.5 mg/L) (Figure 2).

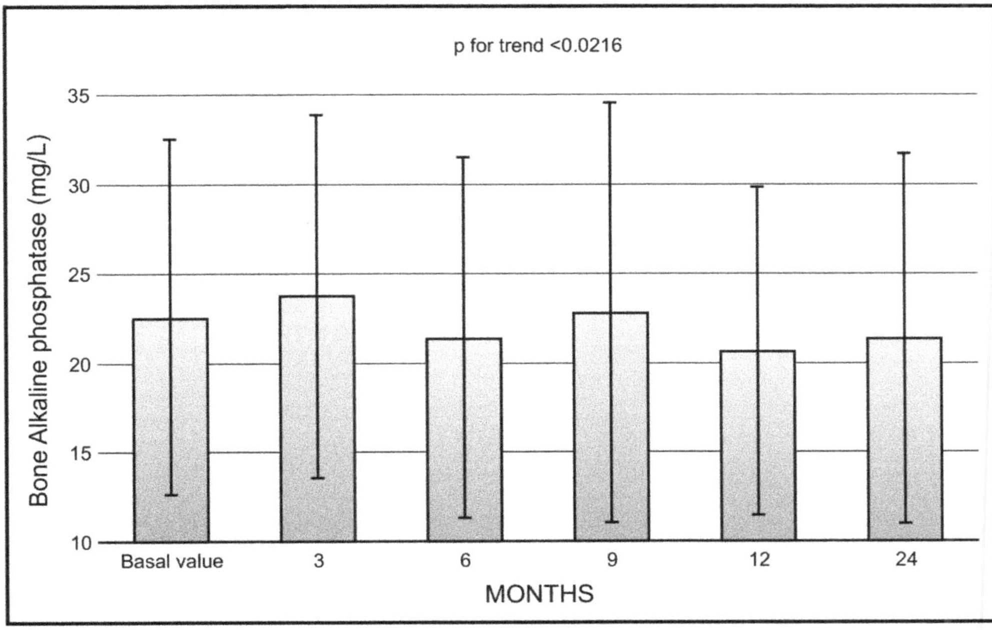

Figure 2. Time course of serum bone alkaline phosphatase following monthly administration of 0.266 mg calcifediol. Error bars indicate 1 SD.

No significant changes were observed as far as mean values of serum ionised calcium (Figure 3) or creatinine were concerned. No serum ionised calcium above 1.33 nmol/L was documented.

Throughout the entire period of treatment, 12 of the patients enrolled reported at least one adverse event (A. E.). Only one treatment-related A. E. was reported by one patient (mild dyspepsia) that in any case did not force her to withdraw from the investigation. Two patients developed a COVID-19 infection during the study. No other safety issues were associated with the present analysis; in particular, there were no deaths and no serious adverse events attributable to calcifediol. The maximum 25(OH)D value reached in the entire population considered for the whole study period was 79.7 ng/mL at month 24, in a patient whose basal value was 26.3 ng/mL.

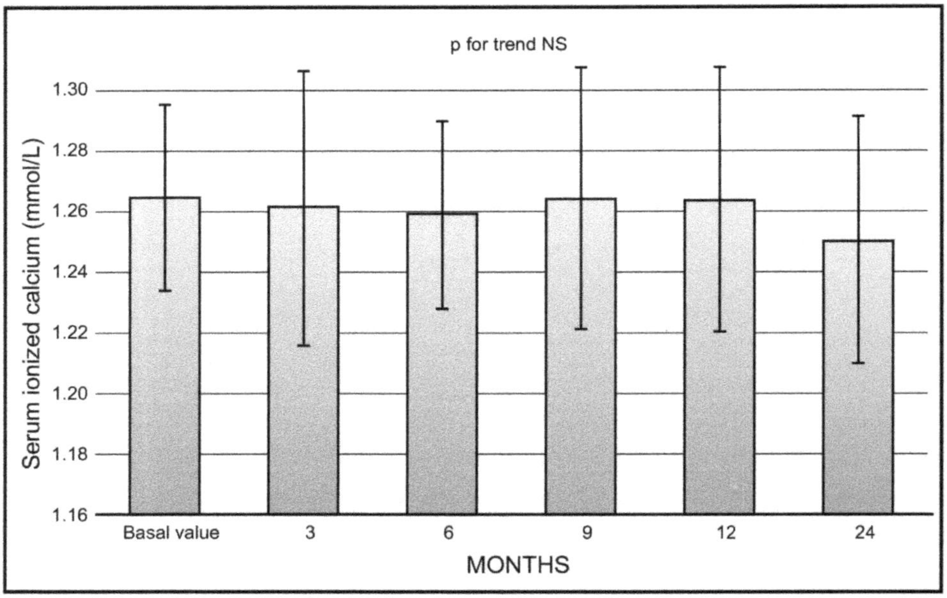

Figure 3. Time course of serum ionised calcium following monthly administration of 0.266 mg calcifediol. Error bars indicate 1 SD.

4. Discussion

To the best of our knowledge, this is the first open-label investigation carried out to evaluate the efficacy and safety of monthly 0.266 mg calcifediol administered for as long as 2 years. Indeed, no previous investigation in the English-language literature has covered such a long period of time with a monthly or other schedule of calcifediol administration. For example, Bischoff-Ferrari and coworkers [4] performed an investigation in 10 healthy postmenopausal women by administering 20 mcg of calcifediol per day over 4 months. They showed that this amount and schedule of administration resulted in a safe, immediate and sustained increase in 25(OH)D serum levels in all participants. Russo and coworkers [12] investigated the effect of the monthly oral administration of 500 mcg of calcifediol in 18 normal women aged 24–72 years for a period of four months. They found that the pulsed monthly administration of 500 mcg of calcifediol can be considered an alternative for vitamin D repletion in respect to cholecalciferol, especially in conditions in which a rapid normalisation of 25(OH)D levels is needed. Graeff-Armas and coworkers [18] administered three different calcifediol doses (10, 15 and 20 mcg daily) to men and postmenopausal women aged 50 years or older, for six months. They essentially demonstrated that 25(OH)D_3 is very effective in raising 25(OH)D concentrations. Furthermore, they showed that once supplementation is discontinued, the elimination rate of 25(OH)D_3 is faster than following the suspension of cholecalciferol, similarly to what has been subsequently reported by Perez-Castrillon and coworkers [14]. Minisola and coworkers [13] enrolled 87 Caucasian, community-dwelling, postmenopausal women, aged 55 years or older, with vitamin D levels below 30 ng/mL. These subjects were randomised to receive three different dosages of calcifediol: 20 mcg daily, 40 mcg daily or 125 mcg weekly for 3 months. They demonstrated that different daily or weekly dosages of calcifediol have a mid-term efficacy and safety on the main parameters of mineral metabolism, when used to treat vitamin D inadequacy or deficiency in this category of subjects. Navarro Valverde et al. enrolled 30 postmenopausal osteoporotic women with an average age of 67 years who were vitamin D-deficient. Calcifediol was administered to three groups of 10 people each, at a dose of 20 mcg daily, 266 mcg weekly or every other week, respectively, for 12 months. At the end of the study, levels above 20 ng/mL were attained by all the subjects, independently of the schedule of administration [19]. Shieh and coworkers [20] enrolled 19 subjects (mean age 34.8 ± 8.6 years), to whom they administered 20 mcg calcifediol daily. They found that calcifediol rapidly and robustly increases serum concentrations of both total and free 25(OH)D and more reliably increases circulating 25(OH)D to levels of 30 ng/mL by the fourth week of supplementation. Gonnelli and coworkers [21] randomised fifty osteopenic/osteoporotic women with serum levels of 25-hydroxyvitamin D between 10 and 20 ng/mL to a 6-month treatment with oral calcifediol 20 mcg/day (n = 23, mean age 62.4 ± 7.4 years) or oral calcifediol 30 mcg/day (n = 25, mean age: 61.5 ± 8.3 years). Calcifediol was able to rapidly normalise the vitamin D deficiency; furthermore, they suggested that the 30 mcg daily dose could be utilised in those patients who need to rapidly reach optimal 25OHD levels. Ruggiero and coworkers administered to community-dwelling women and men, older than 75 years, who were consecutively admitted to a geriatric acute care ward, a weekly dose of 150 mcg of calcifediol. They found that the administration of calcifediol is beneficial in oldest-old people since it rapidly increased serum levels and reaches the optimal target threshold, especially among those with a comorbidity, taking multiple drugs and showing low muscle strength [22]. Finally, Pérez-Castrillòn and collaborators [14] administered 0.266 mg monthly calcifediol for 12 months to 58 postmenopausal women (aged 64.3 ± 8.2 years) with 25(OH)D levels lower than 20 ng/mL. Treatment for one year produced stable and sustained 25(OH)D concentrations with no associated safety concerns.

Fron a review of the previous literature, it is apparent that our investigation fills a gap. Indeed, the 0.266 mg monthly dose of calcifediol has never been previously utilised, with only one notable exception. However, the investigation of Pérez-Castrillòn and coworkers did not have the same duration as the one we carried out. Furthermore, it is also apparent that, unless the reason underlying vitamin D insufficiency is removed, treatment to correct a condition of hypovitaminosis D should be lifelong. Having carried out a study of such a long duration (i.e., 24 months) represents another step forward.

Our data are in line with previous investigations, in that they show the favourable and quick effect of calcifediol in restoring 25(OH)D values to the threshold of sufficiency, regardless of how this is settled by various scientific societies. In this respect, it is important to highlight that, considering the time course of 25(OH)D values following monthly calcifediol administration (Figure 1), starting from the third month of observation, all mean values reached the threshold of 30 ng/mL. The time points of our protocol did not include earlier serum 25(OH)D measurements. However, we previously showed that serum 25(OH)D values significantly and promptly rose and plateaued above the 30 ng/mL threshold, remaining within the safety interval after 14 days of treatment, with a similar efficacy for daily and weekly dose regimens. The different dosages were also equally effective in controlling secondary hyperparathyroidism.

There are some clinical conditions in which a rapid normalisation of 25(OH)D levels is desirable. In this context, it has been shown the low levels of vitamin D might be responsible for the acute-phase reaction induced by the intravenous infusion of amino bisphosphonate and, rarely, also, when it is administered by the oral route. Bertoldo and coworkers showed that patients with 25(OH)D levels below 30 ng/mL have a threefold increased risk of acute-phase reaction associated with zoledronic acid infusion [23]. Even though the mechanisms by which vitamin D status can influence the appearance of acute-phase reaction have not been completely elucidated until now, the most widely accepted mechanism underscores a direct effect on delta $\gamma\delta$ T cells. Indeed, this is the population of T cells mainly involved in the acute-phase reaction. The attainment of vitamin D sufficiency by calcifediol administration in a short period of time is an appealing strategy to avoid this untoward side effect. In fact, it has been proven that the mean increase in serum 25(OH)D after the intake of 1 mcg cholecalciferol is 1.53 nmol/L, whereas the mean increase was 4.76 nmol/L with calcifediol [24]. Next, it has been shown that, in order that drugs utilised for osteoporosis fully exert their effect, patients should be vitamin D replete [25]. We now have available to us a number of drugs (mainly anabolic) that exert their effect in the first few months after initial treatment, as evidenced by Kaplan–Meier estimates in respect to placebo or antiresorptive agents. This rapid effect is especially important in those patients who have sustained a recent fracture. Therefore, a condition of vitamin D sufficiency is a prerequisite to avoid the deleterious effects of secondary hyperparathyroidism. Finally, adequate intakes of calcium and vitamin D are recommended when taking drugs (i.e., denosumab) that heavily suppress osteoclast activity, thus limiting the body's ability to mobilise calcium from bone. In these conditions, the administration of vitamin D with a rapid effect on calcium absorption is mandatory, in order to reduce the risk of treatment-induced hypocalcaemia. In all previous clinical conditions, the administration of calcifediol can be envisaged as a good strategy to allow patients to start treatment immediately. Indeed, the administration of calcifediol has predictable effects on the increase in metabolite values not being dependent on the dose administered. Then, epigenetic factors have a minor role, because the metabolite we measure is the one we administer.

The most important finding of our investigation is represented by the long-term stabilisation of 25(OH)D levels during the whole time frame of our investigation (i.e., 2 years). Indeed, while there was a trend towards increase in the first year of treatment, the levels then plateaued, so that there were no statistically significant differences between mean values obtained at 12 and 24 months of observation. Our results mirror those of previous authors using similar amounts and schedules of calcifediol administration [14]. They showed persistent stable 25(OH)D levels, after reaching the maximum value at 4 months, for the remaining period of 8 months (however, the full length of the study was only one year). Our findings also suggest that, after reaching a steady state, there is no further increase in 25(OH)D values, most probably due to a fine-tuning of CYP24A1. This enzyme is of major clinical and physiologic importance, serving to regulate the catabolism of both 25(OH)D and 1,25(OH)$_2$D, thus avoiding excessive production of the final metabolite of the vitamin D biosynthetic pathway.

We also demonstrated a significant trend toward a decrease in mean values of isoenzymes of alkaline phosphatase, even though, also in this case, the mean values at 12- and 24-month observation were not significantly different. This finding can be interpreted in the light of the inhibitory action of calcifediol on parathyroid hormone secretion, with a consequent decrease in bone turnover. These findings are in line with our previous observations, in which patients were treated with a monthly administration of 500 mcg of calcifediol for a period of 4 months [12]. Unfortunately, not in all studies carried out so far to evaluate the efficacy of calcifediol have markers of bone turnover been assayed. Future studies with measurement of both formation and resorption markers are needed to fill this gap.

Toxicity is an infrequent event in all the studies published so far with calcifediol. In the context of safety, it should be highlighted that safety margins are quite broad, independently of whether we utilise cholecalciferol or its derivatives. In addition, vitamin D toxicity with the classical biochemical (i.e., hypercalcaemia, hypercalciuria, decrease in glomerular filtration rate) and biochemical (i.e., nephrocalcinosis) manifestations is a rare finding, when patients adhere to the prescribed dose of the compound. Most published cases of vitamin intoxication so far reported in the literature indeed arise from self-administering huge doses of vitamin D. It is reassuring that no blood levels above 150 ng/mL have been reported in previous investigations. According to these previous findings, the maximum value of 25(OH)D detected in our investigation was 79.7 ng/mL. It can be hypothesised that extremely high levels of 25(OH)D can be reached if there is a concurrent clinical condition, such as an associated disease that increases the synthesis of 1,25-dihydroxycholecalciferol or genetic defects that alter the metabolism of vitamin D by reducing the concentration of 24,25-dihydroxycholecalciferol, a safety mechanism that prevents toxicity. The main biochemical manifestations of toxicity are represented by the finding of hypercalcaemia and hypercalciuria, together with reduced kidney function. It is worthwhile to emphasise that we did not detect any episode of hypercalcaemia during the entire observation period. This finding, together with the lack of significant changes of serum creatinine and no reported side effects, argue in favour of the safety and tolerability of monthly calcifediol administration. In the biggest randomised controlled trial [14], comparing the efficacy and safety of calcifediol 0.266 mg soft capsules with cholecalciferol in 303 postmenopausal women, no deaths or serious adverse events were reported. Furthermore, the maximum 25(OH) D level reached by a patient was 64.4 ng/mL during the first month of treatment due to a medication error (a weekly instead of monthly intake of calcifediol). The withdrawal criterion for 25(OH)D levels in this clinical trial was a value greater than 80 ng/mL; it is important to highlight that no patient reached that threshold. Also in that study, no clinically relevant hypercalcaemia cases were reported in any of the patients treated with calcifediol. Finally, the maximum calcium level reached was 11.1 mg/dL in one patient, whereas the remaining patients had levels lower than 10.8 mg/dL. Taken together, these data indicate that the finding of an increased serum calcium value is infrequent with a correct administration of 0.266 mg monthly of calcifediol. In our investigation, we utilised a monthly dose of 0.266 mg of calcifediol. However, we are aware of the possibility that such a dose can be increased in specific conditions (i.e., when initially treating a severely deficient patient with vitamin D values less than 12 ng/mL) or when higher thresholds may be desirable (i.e., when treating autoimmune diseases such as rheumatoid arthritis). In these previous conditions, more frequent laboratory tests are advisable, both to check threshold achievement and also the lack of biochemical measures of toxicity. Theoretically, a fine-tuning of the amount of vitamin D to be administered can be envisaged if basal 25(OH)D values are available.

We did not report urinary calcium excretion, because the patients enrolled were not willing to have a troublesome urinary collection in the middle of the COVID-19 pandemic. Urinary collection was performed in a few patients, which did not allow us to have the statistical power to draw meaningful conclusions.

Several limitations of this study should be considered. The first one is represented by the per-protocol, and not intentional, analysis of the data obtained. However, this approach was justified by the peculiar period in which this investigation was carried out (i.e., in the midst of the COVID-19 pandemic) that did not allow a full adherence to the initial scheduled visits of the protocol. In this context should also be interpreted the lack of the inclusion of a placebo control group. Secondly, a full evaluation of the biochemical parameters of skeletal metabolism would have been desirable. However, our focus was the attainment of 25(OH)D values in the sufficiency range in the long term without reporting any serious adverse events. This primary endpoint, not previously targeted in previous investigations, was reached in our study. Finally, we did not include a group treated with cholecalciferol, since there is an abundance of papers showing the different metabolism and predictability of efficacy of calcifediol in respect to cholecalciferol, so that a further study addressing this problem would have resulted in useless repetition.

The main strength of our protocol derives from the homogeneity of the population studied from a single centre and, most importantly, the duration of the study—that is, 24 months—never reached in previous studies published in the English-language literature. Another particular aspect that should be emphasised is represented by the formulation of calcifediol, together with the schedule utilised. Taking one soft gelatine capsule per month is easier compared to taking a certain number of drops each day, as reported in the majority of studies previously cited. This kind of formulation is available both in Italy and Spain, even though recommendations concerning doses may differ (based on initial vitamin D values in Italy or on the characteristics of the population in Spain). The desired final effect on vitamin D status is reached by utilising this formulation, as can be observed by the stable circulating levels of 25(OH)D throughout the whole observation period.

5. Conclusions

In conclusion, this is the first open-label investigation carried out for as long as 2 years with monthly 0.266 mg calcifediol administration. The long-term administration of calcifediol maintained stable and sustained 25(OH)D concentrations. Most importantly, no safety concerns were reported by the patients.

Author Contributions: M.O., J.P., L.C. and C.C., contributed to data collection and analysis, and the drafting and revision of the manuscript; J.P., L.C., V.D.M., M.L., A.A., S.M. and C.C. contributed to data collection and revision of the manuscript; L.N. contributed to data analysis and revision of the manuscript and carried out the statistical analysis; S.M. contributed to the study design. All authors have read and agreed to the published version of the manuscript.

Funding: This study was partly funded by "Progetti di Ricerca—Ateneo 2018" Sapienza University.

Institutional Review Board Statement: This study was conducted according to the guidelines of the Declaration of Helsinki and approved by the Ethics Committee of Policlinico Umberto I (protocol code 3267, 26 June 2014, amended 6828, 7 December 2022).

Informed Consent Statement: Informed consent was obtained from all subjects involved in the study.

Data Availability Statement: Some or all datasets generated during the current study are not publicly available due to privacy but are available from the corresponding author upon reasonable request.

Conflicts of Interest: S.M. served as a speaker for Abiogen, Beijing Society Biomedical Engineering. Geopharma, Sandoz, UCB. He also served on the advisory board of Kyowa Kirin and Faes Farma. C.C. served as speaker for Abiogen and on the advisory board of IBSA. The other authors have no conflicts to declare.

References

1. Minisola, S.; Colangelo, L.; Pepe, J.; Diacinti, D.; Cipriani, C.; Rao, S.D. Osteomalacia and vitamin D status: A clinical updat *JBMR Plus* **2020**, *5*, e10447. [CrossRef] [PubMed]
2. Chevalley, T.; Brandi, M.L.; Cashman, K.D.; Cavalier, E.; Harvey, N.C.; Maggi, S.; Cooper, C.; Al-Daghri, N.; Bock, O.; Bruyère, C et al. Role of vitamin D supplementation in the management of musculoskeletal diseases: Update from an European Society c Clinical and Economical Aspects of Osteoporosis, Osteoarthritis and Musculoskeletal Diseases (ESCEO) working group. *Agin Clin. Experim Res.* **2022**, *34*, 2603–2623. [CrossRef] [PubMed]
3. Bouillon, R.; Manousaki, D.; Rosen, C.; Trajanoska, K.; Rivadeneira, F.; Richards, J.B. The health effects of vitamin D supplement tion: Evidence from human studies. *Nat. Rev. Endocrinol.* **2020**, *18*, 96–110. [CrossRef] [PubMed]
4. Bischoff-Ferrari, H.A.; Dawson-Hughes, B.; Stocklin, E.; Sidelnikov, E.; Willett, W.C.; Edel, J.O.; Stahelin, H.B.; Wolfram, S.; Jette A.; Schwager, J.; et al. Oral supplementation with 25(OH)D3 versus vitamin D3: Effects on 25(OH)D levels, lower extremit function, blood pressure, and markers of innate immunity. *J. Bone Miner. Res.* **2012**, *27*, 160–169. [CrossRef] [PubMed]
5. Akutsu, T.K.H.; Himeiwa, S.; Kitada, S.; Akasu, T.; Urashima, M. Vitamin D and cancer survival: Does vitamin D supplementatio improve the survival of patients with cancer? *Curr. Oncol. Rep.* **2020**, *22*, 62. [CrossRef] [PubMed]
6. Sutherland, J.P.; Zhou, A.; Hyppönen, E. Vitamin D deficiency increases mortality risk in the UK Biobank: A nonlinear mendelia randomization study. *Ann. Intern. Med.* **2022**, *175*, 1552–1559. [CrossRef] [PubMed]
7. Bouillon, R.; Quesada Gomez, J.M. Comparison of calcifediol with vitamin D for prevention or cure of vitamin D deficienc *J. Steroid Biochem. Mol. Biol.* **2023**, *228*, 106248. [CrossRef] [PubMed]
8. Ebeling, P.R.; Adler, R.A.; Jones, G.; Liberman, U.A.; Mazziotti, G.; Minisola, S.; Munns, C.F.; Napoli, N.; Pittas, A.G.; Giustin A.; et al. Management of Endocrine Disease: Therapeutics of Vitamin D. *Eur. J. Endocrinol.* **2018**, *179*, R239–R259. [CrossRe [PubMed]
9. Bishop, C.W.; Strugnell, S.A.; Csomor, P.; Kaiser, E.; Ashfaq, A. Extended-release calcifediol effectively raises serum total 2 hydroxyvitamin D even in overweight nondialysis chronic kidney disease patients with secondary hyperparathyroidism. *Am. Nephrol.* **2022**, *53*, 446–454. [CrossRef]
10. Minisola, S.; De Martino, V. Disentanglement among vitamins D. *Gynecol. Endocrinol.* **2022**, *28*, 97–98. [CrossRef]
11. Charoenngam, N.; Kalajian, T.A.; Shirvani, A.; Yoon, G.H.; Desai, S.; McCarthy, A.; Apovian, C.M.; Holick, M.F. A pilo randomized, double-blind crossover trial to evaluate the pharmacokinetics of orally administered 25-hydroxyvitamin D3 an vitamin D3 in healthy adults with differing BMI and in adults with intestinal malabsorption. *Am. J. Clin. Nutr.* **2021**, *11* 1189–1199. [CrossRef] [PubMed]
12. Russo, S.; Carlucci, L.; Cipriani, C.; Ragno, A.; Piemonte, S.; Fiacco, R.D.; Pepe, J.; Fassino, V.; Arima, S.; Romagnoli, E.; et a Metabolic changes following 500 mug monthly administration of calcidiol: A study in normal females. *Calcif. Tissue Int.* **2011**, *8* 252–257. [CrossRef] [PubMed]

13. Minisola, S.; Colangelo, L.; Biondi, P.; Cipriani, C.; Fossi, C.; Franceschelli, F.; Giusti, F.; Leoncini, G.; Pepe, J.; Bischoff-Ferrari, H.A.; et al. Correction of vitamin D status by calcidiol: Pharmacokinetic profile, safety, and biochemical effects on bone and mineral metabolism of daily and weekly dosage regimens. *Osteop Int.* **2017**, *28*, 3239–3249. [CrossRef] [PubMed]
14. Pérez-Castrillón, J.L.; Dueñas-Laita, A.; Gómez-Alonso, C.; Jódar, E.; del Pino-Montes, J.; Brandi, M.L.; Cereto Castro, F.; Quesada-Gómez, J.M.; Gallego López, L.; Olmos Martínez, J.M.; et al. Long-term teatment and effect of discontinuation of calcifediol in postmenopausal women with vitamin D deficiency: A randomized trial. *J. Bone Miner. Res.* **2023**, *38*, 471–479. [CrossRef] [PubMed]
15. Cipriani, C.; Minisola, S.; Bilezikian, J.P.; Diacinti, D.; Colangelo, L.; Piazzolla, V.; Angelozzi, M.; Nieddu, L.; Pepe, J.; Diacinti, D. Vertebral Fracture Assessment in postmenopausal women with postsurgical hypoparathyroidism. *J. Clin. Endocrinol. Metab.* **2021**, *106*, 1303–1311. [CrossRef] [PubMed]
16. Danese, V.C.; Pepe, J.; Ferrone, F.; Colangelo, L.; De Martino, V.; Nieddu, L.; Ferrazza, G.; Panzini, E.; Pascone, R.; Blocki, F.; et al. The mutual interplay between bone, glucose and lipid metabolism: The role of vitamin D and PTH. *Nutrients* **2023**, *15*, 2998. [CrossRef] [PubMed]
17. Pedrazzoni, M.; Girasole, G.; Bertoldo, F.; Bianchi, G.; Cepollaro, C.; Del Puente, A.; Giannini, S.; Gonnelli, S.; Maggio, D.; Marcocci, C.; et al. Definition of a population-specific DXA reference standard in Italian women: The Densitometric Italian Normative Study (DINS). *Osteoporos. Int.* **2003**, *14*, 978–982. [CrossRef] [PubMed]
18. Graeff-Armas, L.A.; Bendik, I.; Kunz, I.; Schoop, R.; Hull, S.; Beck, M. Supplemental 25-hydroxycholecalciferol is more effective than cholecalciferol in raising serum 25-hydroxyvitamin D concentrations in older adults. *J. Nutr.* **2020**, *150*, 73–81. [CrossRef] [PubMed]
19. Navarro-Valverde, C.; Sosa-Henríquez, M.; Alhambra-Expósito, M.R.; Quesada-Gómez, J.M. Vitamin D3 and calcidiol are not equipotent. *J. Steroid Biochem. Mol. Biol.* **2016**, *164*, 205–208. [CrossRef]
20. Shieh, A.; Ma, C.; Chun, R.F.; Witzel, S.; Rafison, B.; Contreras, H.T.M.; Wittwer-Schegg, J.; Swinkels, L.; Huijs, T.; Hewison, M.; et al. Effects of cholecalciferol vs calcifediol on total and free 25-hydroxyvitamin D and parathyroid hormone. *J. Clin. Endocrinol. Metab.* **2017**, *102*, 1133–1140. [CrossRef]
21. Gonnelli, S.; Tomai Pitinca, M.D.; Camarri, S.; Lucani, B.; Franci, B.; Nuti, R.; Caffarelli, C. Pharmacokinetic profile and effect on bone markers and muscle strength of two daily dosage regimens of calcifediol in osteopenic/osteoporotic postmenopausal women. *Aging Clin. Experim Res.* **2021**, *33*, 2539–2547. [CrossRef] [PubMed]
22. Ruggiero, C.; Baroni, M.; Bini, V.; Brozzetti, A.; Parretti, L.; Zengarini, E.; Lapenna, M.; Antinolfi, P.; Falorni, A.; Mecocci, P.; et al. Effects of weekly supplementation of cholecalciferol and calcifediol among the oldest-old people: Findings from a randomized pragmatic clinical trial. *Nutrients* **2019**, *11*, nu11112778. [CrossRef] [PubMed]
23. Bertoldo, F.; Pancheri, S.; Zenari, S.; Boldini, S.; Giovanazzi, B.; Zanatta, M.; Valenti, M.T.; Dalle Carbonare, L.; Lo Cascio, V. Serum 25-hydroxyvitamin D levels modulate the acute-phase response associated with the first nitrogen-containing bisphosphonate infusion. *J. Bone Miner. Res.* **2010**, *25*, 447–454. [CrossRef] [PubMed]
24. Quesada-Gomez, J.M.; Bouillon, R. Is calcifediol better than cholecalciferol for vitamin D supplementation? *Osteop. Int.* **2018**, *29*, 1697–1711. [CrossRef]
25. Adami, S.; Giannini, S.; Bianchi, G.; Sinigaglia, L.; Di Munno, O.; Fiore, C.E.; Minisola, S.; Rossini, M. Vitamin D status and response to treatment in post-menopausal osteoporosis. *Osteop. Int.* **2009**, *20*, 239–244. [CrossRef]

Disclaimer/Publisher's Note: The statements, opinions and data contained in all publications are solely those of the individual author(s) and contributor(s) and not of MDPI and/or the editor(s). MDPI and/or the editor(s) disclaim responsibility for any injury to people or property resulting from any ideas, methods, instructions or products referred to in the content.

nutrients

MDPI

Review

Nutrition, Vitamin D, and Calcium in Elderly Patients before and after a Hip Fracture and Their Impact on the Musculoskeletal System: A Narrative Review

Luisella Cianferotti [1,*], Giuseppe Bifolco [1], Carla Caffarelli [2], Gherardo Mazziotti [3], Silvia Migliaccio [4], Nicola Napoli [5], Carmelinda Ruggiero [6] and Cristiana Cipriani [7]

[1] Bone Metabolic Diseases Unit, Department of Experimental and Clinical Biomedical Sciences, University Hospital of Florence, University of Florence, 50134 Florence, Italy; giuseppe.bifolco@unifi.it
[2] Division of Internal Medicine, Department of Medicine, Surgery and Neuroscience, University of Siena, 53100 Siena, Italy; carla.caffarelli@unisi.it
[3] Department of Biomedical Sciences, Humanitas University, Endocrinology, Diabetology and Andrology Unit, IRCCS Humanitas Research Hospital, 20089 Milan, Italy; gherardo.mazziotti@hunimed.eu
[4] Department of Experimental Medicine, University Sapienza of Rome, 00185 Rome, Italy; silvia.migliaccio@uniroma1.it
[5] Unit of Endocrinology and Diabetes, Department of Medicine, Rome Biomedical Campus University Foundation, 00128 Rome, Italy; n.napoli@unicampus.it
[6] Geriatric and Orthogeriatric Units, Division Gerontology and Geriatrics, Department of Medicine and Surgery, University of Perugia, 06156 Perugia, Italy; carmelinda.ruggiero@unipg.it
[7] Department of Clinical, Internal, Anesthesiological and Cardiovascular Sciences, Sapienza University of Rome, 00185 Rome, Italy; cristiana.cipriani@uniroma1.it
* Correspondence: luisella.cianferotti@unifi.it

Abstract: Hip fractures are a major health issue considerably impacting patients' quality of life and well-being. This is particularly evident in elderly subjects, in which the decline in bone and muscle mass coexists and predisposes individuals to fall and fracture. Among interventions to be implemented in hip fractured patients, the assessment and management of nutritional status is pivotal, particularly in subjects older than 65. Nutrition plays a central role in both primary and secondary preventions of fracture. An adequate protein intake improves muscle mass and strength and the intestinal absorption of calcium. Other nutrients with recognized beneficial effects on bone health are calcium, vitamins D, K, and C, potassium, magnesium, folate, and carotenoids. With reference to calcium, results from longitudinal studies showed that the consumption of dairy foods has a protective role against fractures. Moreover, the most recent systematic reviews and meta-analyses and one umbrella review demonstrated that the combination of calcium and vitamin D supplementation significantly reduces hip fracture risk, with presumed higher efficacy in older and institutionalized subjects. Owing to these reasons, the adequate intake of calcium, vitamin D, protein, and other macro and micronutrients has been successfully implemented in the Fracture Liaison Services (FLSs) that represent the most reliable model of management for hip fracture patients. In this narrative review, papers (randomized controlled trials, prospective and intervention studies, and systematic reviews) retrieved by records from three different databases (PubMed, Embase, and Medline) have been analyzed, and the available information on the screening, assessment, and management of nutritional and vitamin D status and calcium intake in patients with hip fractures is presented along with specific prevention and treatment measures.

Keywords: osteoporosis; fragility fracture; falls; cholecalciferol; calcium carbonate; calcium citrate; calcium intake

1. Introduction

The prevalence of osteoporosis and related fragility fractures is steadily rising worldwide [1]. Among fragility fractures, hip fractures represent a major health concern, especially in the aging population, namely over the age of 65, and are often the result of falls [2]. Hip fracture is an often-devastating event, likely leading to a hampered quality of life, disability, associated complications, impact on a person's overall well-being, and increased risk of mortality [3]. Preventative measures and effective treatment are essential for reducing the incidence of these fractures, preventing further fractures, and improving quality of life for those who experience them [4]. Understanding the epidemiology and consequences of hip fractures is essential for healthcare planning, preventive measures, and addressing the specific needs of populations at higher risk.

Public health initiatives and policies aimed at reducing the incidence of hip fractures have to target osteoporosis prevention, fall prevention, and improve the overall well-being of the elderly population. Once a major fragility fracture such a hip fracture has occurred, proper treatment with antiosteoporotic agents becomes mandatory as a secondary prevention strategy for further fractures, the risk of which is imminently high right in the first two years after the first event [5]. Unfortunately, the minority of people who experience a hip fracture are evaluated and begin proper therapy. The treatment gap in osteoporosis refers to the significant discrepancy between the number of people who have osteoporosis or are at high risk of fractures and those who actually receive appropriate diagnosis, treatment, and management for their condition [6]. This gap is a major concern in healthcare for several reasons, contributing to increased rates of fractures, higher healthcare costs, and decreased quality of life for those affected.

Regarding the secondary prevention of fractures, Fracture Liaison Services (FLSs), primarily borne within tertiary clinical centers around the patient with a hip fracture, represent coordinated, multidisciplinary systems designed to identify, diagnose, treat, and manage the care of patients with osteoporotic fractures, with the ultimate goal of preventing subsequent fractures [7]. As a critical component of integrated care for patients suffering from fractures, especially those at risk of osteoporosis or subsequent fractures, the FLS plays a pivotal role in addressing what is often an under-recognized and under-treated condition. Despite the clear evidence of the cost-effectiveness of the FLSs where applied, they are not routinely supported with specific resources by healthcare systems.

Nutrition plays a pivotal role in the health and functionality of the musculoskeletal system in elderly individuals, directly impacting their mobility, strength, and risk of falls [8]. The derangement from aging to frailty is often accompanied by an accelerated decline in muscle mass and bone density, leading to a pathological condition known as sarco-osteoporosis. These changes can significantly increase the risk of falls, fractures, and subsequent disabilities, highlighting the importance of nutritional and exercise strategies to mitigate these risks. In particular, nutrition plays a pivotal role in the primary and secondary prevention of osteoporosis and fragility fractures [9] (Figure 1). A good nutritional condition is correlated to better musculoskeletal health in all ages, along with a decrease in the probability of falls and subsequent fractures [9] (Figure 1). Nutritional status is often compromised in elderly people, so that the typical patient with a hip fracture, the risk of which exponentially rises after 65 years of age in both sexes with more than 75% occurring in people older than 75 years, is often nutritionally impaired.

Nutritional status can be evaluated by general nutritional questionnaires, especially formulated for elderly people, as well as vitamin D status and calcium intake assessments. These factors are often neglected in the majority of patients, especially in aged patients after a fragility fracture, so that both patients and care givers are not usually sufficiently educated on the importance of a proper diet and how to manage to achieve it.

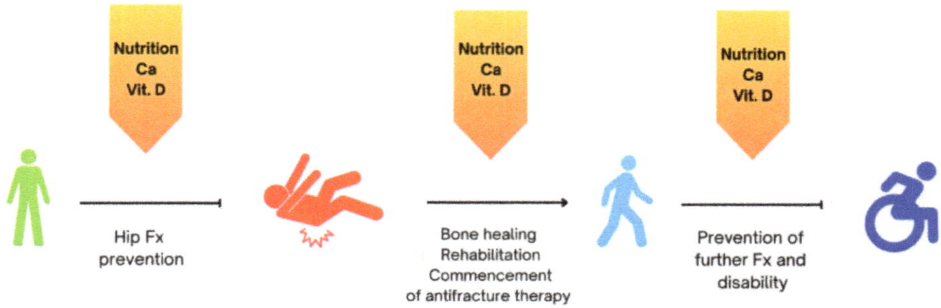

Figure 1. The optimization of nutrition, calcium intake, and vitamin D status is a multimodal approach which can intervene at various stages in the prevention of falls and related hip fragility fractures, recovery from hip fracture, and the prevention of further fractures, along with proper antifracture drugs and related disability.

The correction of vitamin D status by vitamin D supplementation, even besides baseline vitamin D levels as assessed at the moment of fracture, is mandatory in patients after a major fragility fracture, in order to commence proper treatment early [9]. Indeed, despite the worldwide increase in the consumption of vitamin D supplements in the general population, their use remains low in patients experiencing fragility fractures and, in particular, at the moment of a hip fracture. Mineral intake, and calcium intake in particular, is fundamental for proper bone repair, to optimize the mineralization process and to favor the action of anti-osteoporotic medications. In addition, it has been proven that adverse events, such as hypocalcemia, common for most potent antiresorptives, are less frequent when calcium ingestion is adequate. Calcium intake, which would have always to be optimized in parallel to the improvement in vitamin D status, is usually not systematically evaluated by specific questionnaires, nor have specific interventions been applied systematically in the general management of these patients.

In this review, we will explore the state of the art on the assessment of calcium intake, nutrition, and vitamin D status in patients before and after a hip fracture and analyze whether systematic specific interventions have been undertaken to optimize these parameters in these individuals.

2. Methods

For this narrative review, data from three different databases (PubMed, Embase and Medline) were retrieved using the following key words: "hip fracture", "osteoporosis", "vitamin D", "serum 25(OH)D", "calcium intake", "nutrition", "malnutrition", "protein intake", "falls", "disability", "mortality", "vitamin D status", "calcium intake questionnaire", "elderly", and "elders", using "AND"/"OR" operators as appropriate.

Randomized controlled trials (RCTs), prospective and intervention studies, systematic reviews, and meta-analyses in the English language, mainly focusing on the literature of the last 10 years up to December 2023, were considered.

Abstracts were reviewed by 3 reviewers (CC, BG, and CL) and only main papers and reviews referring to the prevention of hip fractures and the assessment/treatment of patients with hip fractures and published in English were considered.

3. Nutrition and Hip Fracture

In elderly people, a number of physiological factors, such as a loss of appetite, chewing problems, dysphagia, reduced mobility, and psychological problems, and environmental factors, such as mourning, social isolation, and economic problems, can influence eating habits with advancing age [10–12].

People's overall well-being is closely related to nutritional status. Regarding musculoskeletal health, it is crucial to ensure in the elderly an adequate protein intake, as proteins are involved in bone mineral metabolism and are able to improve bone matrix protein collagen [13]. The European Society for Clinical Nutrition and Metabolism (ESPEN) recommended for older adults with acute or chronic illnesses 1.2 to 1.5 g protein/kg body weight/day, increasing the intake in the case of more serious illnesses [14] in order to avoid a progressive loss of lean mass, determined by a nitrogen imbalance [15]. In addition, it appears that an increased protein intake may promote the growth of muscle mass and strength in the lower limbs and increase the intestinal absorption of calcium [16,17].

A poor diet can trigger a range of both physical and mental problems, increasing the risk of conditions such as osteoporosis and falls and recurrent falls [18,19]. These difficulties can contribute to malnutrition due to limited food and fluid intake and due to the difficulty of meeting recommended nutritional requirements. Malnutrition per se is widely recognized for its detrimental effects on various physiological systems, which can lead to increased complications after surgery and increased mortality rates [20]. During the aging process, there is a decrease in the intake of macronutrients and micronutrients, and a decrease in the adherence to a healthy diet, which would allow one to maintain bone mineral mass [21]. Usually, a healthy diet is characterized by a high consumption of whole grains, fruits, vegetables, and legumes that are rich in the micronutrients, such as calcium, vitamin K, potassium, magnesium, vitamin C, folate, and carotenoids, necessary for bone health [22]. Several studies have shown that a high adherence to the Mediterranean diet, ensuring an adequate intake of calcium and vitamin D through food, correlates with a low risk of hip fractures in the elderly [23–26]. In contrast, other types of different dietary patterns, such as the Western diet, rich in red meat, processed meat, poultry with skin, animal organ meat, cooking oil, soft drinks, hamburgers, hotdogs, ice cream, doughnuts, margarine, and butter [27], and certain nutrients such as saturated fatty acids and trans fatty acids, have been shown to be associated with higher levels of inflammation. Consequently, the inflammatory component of the diet may increase the risk of hip fractures in the elderly population without a gender difference [28,29].

As for the role of proteins, scientific evidence demonstrates that a high dietary protein intake is important for fall prevention in old people [30] and it is associated with a low risk of hip fractures [31]. Recently, interesting evidence has been found by Hayashi et al. [32], who show that not only total protein intake but a more frequent consumption of meals with adequate amounts of protein (≥20 g or ≥30 g) may represent a potential strategy to oppose the age-related loss of muscle mass and function, especially in frail and pre-fragile individuals.

In the UK's Women's Cohort Study, including 26,318 women aged 35–69 years, with 822 hip fracture episodes retrieved by hospital records in the >20-year follow-up, an increased intake of protein, calcium, total dairy, and milk was associated with a 45% lower risk of hip fracture for every 25 g/day protein consumed in underweight individuals [33]. In a further analysis of the same cohort, vegetarian women displayed higher risks of hip fracture with respect to regular meat eaters [34]. These findings, although not yet confirmed in males and in other non-European populations, highlight the importance of certain diets and proper nutrition with certain type of foods.

On admission to a hospital, nursing home, or during outpatient follow-up, as in the case of people experiencing a hip fracture, it is important to assess the patient's nutritional status. According to the ESPEN guidelines, nutritional status should be diagnosed with standard screening tools, such as the Mini Nutrition Assessment (MNA), Nutrition Risk Screening 2002 (NRS), or the Malnutrition Universal Screening Tool (MUST) [1,35].

Upon admission to the hospital, traditional anthropometric measurements such as weight or body mass index do not allow for the easy identification of the patient.

Therefore, screening malnutrition in older people with simple tools can be critical for early diagnosis and can direct targeted nutritional interventions to improve postoperative outcomes. According to epidemiological data, malnutrition prevalence in hospitalized elderly patients ranges from 20 to 50% [36]. The prevalence of malnutrition is lower, between 2 and 32%, among the elderly residing with family, and increases with age [10,37]. In addition, one year after hip surgery, patients with malnutrition have a significantly increased risk of mortality [38].

Patients admitted to the hospital for a hip fracture showed a low intake of carbohydrates, proteins, fats, certain vitamins, and minerals. The majority of these patients were getting less than the recommended minimum intake, especially for selenium, magnesium, iron, folate, calcium, and zinc [39].

The maintenance of good nutritional status in these patients is necessary in order for them to meet their energy requirements. Oral nutritional supplementation (ONS) is generally used during rehabilitation in hip fracture patients with the aim of increasing energy and protein intake, but no convincing effect on mobility, independence, muscle strength, or muscle function has been observed [40].

Several interventions have been proposed to improve nutrition in hip fracture patients, such as not-otherwise-specified enteral nutrition, which seems to have benefits in decreasing postoperative interleukin-6 levels, which could be related to a reduction in time spent in bed and an improved quality of life [41]. Nonetheless, contradictory results are available on the nutritional support via a nasogastric tube, since this procedure has not shown promising results in these patients [42].

In a multicenter randomized controlled trial (RCT), Wyers et al. monitored patients for three months after a hip fracture, with weekly nutritional counselling and intervening with an energy- and protein-rich diet and using ONS [43]. The study showed no effect on lifespan, postoperative complications, or functional parameters, nor on fracture and mortality rates at 1 and 5 years [43]. Recently, Iuliano et al., in another RCT, tested a nutritional intervention with high-calcium dairy foods to increase calcium and protein intake and calcium in institutionalized elderly people, as detailed later in this paper, demonstrating a reduction in falls, total fractures, and hip fractures [44].

It seems that interventions that just administer supplements are insufficient and unable to improve clinical outcomes [45]. Indeed, in patients with hip fracture, no positive effects were found even by intervening with just taurine supplementation [46], a semi-essential amino acid with antioxidant action in humans. Bell et al. showed that offering individualized, multidisciplinary, multimodal care involving different professionals such as orthopedists, dietitians, and nursing staff can break down the barriers that cause reduced food intake, and is a key strategy for improving clinical outcomes [47]. Involving a nutritionist and nurses with specific expertise in nutrition has been found to correlate with an increased intake of energy, protein, and supplements [48,49]. These studies demonstrate that a multidisciplinary approach is essential for reducing malnutrition in hospitalized patients. Such a multidisciplinary approach was found to be effective in counteracting the increased incidence of malnutrition after discharge.

The elderly also experience recurrent falls, which seems to be associated with fatigue, reduced strength, and muscle quality. Therefore, important strategies that are able to work directly on the muscle and preserve it have the potential to improve results. Significant improvements in function and disability reduction in these patients have been achieved by combining structured exercise programs with essential amino acid supplementation [50]. In a cohort study of community-dwelling patients, these conditions appear to be related to nutritional deficiencies and lower hemoglobin levels [51]. Consistent with low hemoglobin levels, there is a reduction in transported oxygen, leading to a reduction in muscle function and occasional falls.

It seems that one of the strategies to ensure greater patient well-being and improve outcomes is to intervene with a multidisciplinary approach, underlying that fractures are not only an orthopedic issue.

In fact, due to limited access to osteoporosis treatment, several targeted care programs have been designed for patients with this condition and fragility fractures. Programs such as the FLS were implemented to reduce the lack of attention and care for patients with fragility fractures. It seems to result as an effective model; scientific evidence shows that there was increased adherence to medication use, increased exercise, decreased falls, and an increased intake of adequate calcium, vitamin D3, and protein [52,53]. In addition, one-year mortality rates were generally low [54], and the overall re-fracture rate was lower than that reported in previous studies. Currently, the best way to intervene with the dosage of nutritional supplementation, the ideal duration of interventions, and the frequency of nutritional counseling remain unclear. Adherence to the nutritional supplementation regimen and intervention is highly variable, depending on the duration and modality of supplementation [55].

4. Vitamin D Status and Hip Fractures

Hypovitaminosis D is extremely common in hip fracture patients. The overall prevalence of vitamin D deficiency is reported by studies performed in different countries in the range of 46–92%, depending on the series and thresholds of the serum 25(OH)D used to define deficiency [56–67]. Additionally, results from recent systematic reviews and meta-analyses corroborated the hypothesis that low serum 25(OH)D levels represent a risk factor for hip fracture [68–72]. The main results from the meta-analyses published in the last ten years on this topic and on the association between serum 25(OH)D levels and post-fracture outcomes are summarized in Table 1.

Table 1. Summary of the meta-analyses of observational studies assessing the association between serum 25-hydroxy-vitamin D [25(OH)D] levels and hip fracture risk and post-fracture outcomes published from 2014 to 2024.

Author and Year of Publication	Design of Included Studies	Population	Primary Outcome Measure/s	Results
Lv et al., 2017 [68]	Prospective cohort	adults	Association between serum 25(OH)D levels and hip fracture risk	- Significant increased hip fracture risk [RR 1.58 (1.41, 1.77)] in subjects with the lowest vs. the highest serum 25(OH)D levels (15 studies, 51,239 participants); - Subgroup analysis by gender, age, location, type of study, and follow-up confirmed the main results; - The dose–response meta-analysis showed that the increased hip fracture risk is significant when serum 25(OH)D levels are <60 nmol/L.
Feng et al., 2017 [72]	Prospective cohort, case–control	adults	Association between serum 25(OH)D levels and hip fracture risk	- Significant increased hip fracture risk (RR 1.48, 95% CI 1.29–1.68) in subjects with low serum 25(OH)D levels (11 studies); - Subgroup analysis showed 40% increase in hip fracture risk for any SD decrease in serum 25(OH)D levels (RR 1.40, 95% CI 1.20–1.61) (3 studies).
Yao et al., 2019 [73]	Prospective cohort, case–control	adults	Association between serum 25(OH)D levels and hip fracture risk	- Significant 20% decrease in hip fracture risk (RR, 0.80; 95%CI, 0.75–0.86) for any 10 ng/mL increase in serum 25(OH)D levels (11 studies, 39,141 participants); - Subgroup analysis by study design, age, location, follow-up, and baseline 25(OH)D levels showed no significant difference.
Wang et al., 2020 [69]	Prospective cohort	>60 years	Association between serum 25(OH)D levels and hip fracture risk	- Significant increased hip fracture risk [adjusted relative risk (95% CI) 0.89 (0.80, 0.98)] in association with low serum 25(OH)D levels (13 studies, 24,220 participants); - Subgroup analysis by gender, location, and starting time of the study confirmed results for studies performed in Europe, US, and before 2010.

Table 1. *Cont.*

Author and Year of Publication	Design of Included Studies	Population	Primary Outcome Measure/s	Results
Ghahfarrokhi et al., 2022 [70]	Case–control, cohort, retrospective	elderly	Association between serum 25(OH)D levels and hip fracture risk	- Significant increased hip fracture risk [OR 1.80 (95% CI 1.56–2.07, $p \leq 0.001$)] in subjects with low vs. those with high 25(OH)D serum levels (28 studies, 61,744 participants); - Subgroup analysis by study design, gender, sample size, follow-up, study period, NOS score, and geographical location confirmed the main results.
Llombart et al., 2023 [74]	Prospective cohort, retrospective cohort	≥50 years	Association between serum 25(OH)D levels and re-habilitation and functional status in subjects with hip fracture	- No significant relationship between the ability to walk and vitamin D deficiency (OR 0.68, 95% CI 0.31–1.53) (3 studies, 586 participants); - No significant relationship between the length of stay in the hospital and vitamin D deficiency (MD 2.27, 95% CI − 2.47 to 7.01) (3 studies, 1395 participants); - Significant relationship between a worse quality of life and functional ability and vitamin D deficiency (SMD −1.50, 95% CI −2.88 to −0.12) (3 studies, 573 participants).
Llombart et al., 2024 [75]	Prospective cohort, retrospective cohort	adults	Association between serum 25(OH)D levels and mortality in subjects with hip fracture	- Significant increased risk of mortality [OR 1.24 (95% CI 1.05–1.46)] in subjects with vitamin D insufficiency (9 studies, 4409 participants); - Analysis adjusted for age, sex, serum albumin levels, and comorbidities showed no significant association between low 25(OH)D levels and mortality; - Subgroup analysis by follow-up time showed significant increased mortality [OR 1.37 (1.06–1.77) and 1.78 (1.01–3.15)] in subjects with vitamin D insufficiency after 1 and 2 years from hip fracture, respectively.
Wang et al., 2024 [76]	Prospective cohort, retrospective cohort	adults	Association between serum 25(OH)D levels and mortality in subjects with hip fracture	- Significant increased risk of mortality [HR 2.29 (95% CI 1.41–3.72)] in subjects with vitamin D deficiency (14 studies); - No significant increased risk of mortality [HR 1.10 (95% CI 0.97–1.24)] in subjects with vitamin D insufficiency (11 studies); - Subgroup analysis by location confirmed the association between vitamin D deficiency and mortality in studies conducted in Europe [HR 2.4 (95% CI 1.28–4.52)], but not in America and Australia; - Subgroup analysis demonstrated no significant association between vitamin D deficiency and recovery of walking ability [HR 1.77 (95% CI 0.95–3.32)].

RR, relative risk; OR, odds ratio; NOS, Newcastle–Ottawa Scale; MDs, mean differences; SMDs, standardized mean differences; HR, hazard ratio.

Lv et al. reported a pooled relative risk (RR) of hip fracture of 1.58 in association with the lowest vs. the highest serum 25(OH)D levels in a meta-analysis of 13 prospective cohort studies [68]. In particular, subgroup analysis demonstrated that hip fracture risk was increased when low serum 25(OH)D levels were measured in women and subjects older than 65 [68]. Similar results were described by Wang et al. in a meta-analysis of 13 cohort studies in 24,220 patients (2831 with hip fractures) older than 60, in which higher relative risks of hip fracture in association with low 25(OH)D levels were reported, with a RR of 1.25 in female subjects [69]. The analysis of the association between hip fracture risk and the magnitude of serum 25(OH)D decrease showed interesting results, as well. Feng et al.

reported a 40% increase in hip fracture risk for any SD decrease in serum 25(OH)D levels in the subgroup meta-analysis of three prospective cohort and case–control studies [2,72]. Interestingly, Yao et al. described a 20% lower risk of hip fracture for any 10 ng/mL increase in serum 25(OH)D concentration in a meta-analysis of 11 observational studies with 39,141 participants aged 52–76, of whom 2367 had a hip fracture [73]. However, heterogeneity among studies was reported [73]. A more recent systematic review and meta-analysis of 28 studies performed in elderly subjects with a total of 9767 hip fractures clarified that the fracture risk was increased in association with low 25(OH)D serum levels when case–control, cohort, and case–cohort studies from different geographical areas were considered [70].

With reference to post-surgical outcomes, retrospective and prospective studies from several countries collectively showed worse functional recovery in association with lower 25(OH)D concentration [77,78]. A secondary analysis of the Functional Outcomes in Cardiovascular Patients Undergoing Surgical Hip Fracture Repair (FOCUS) Trial including 290 hip fractured subjects 65 and older undergoing surgery between 2004 and 2009 showed reduced mobility in association with vitamin D deficiency before surgery [77]. More specifically, preoperative serum 25(OH)D > 12 ng/mL were associated with 30- and 60-day increased mobility [77]. Subsequent retrospective analyses confirmed the association between pre-surgical vitamin D deficiency and 6- or 12-month worse physical outcomes, re-admission for medical issues, and re-fracture in elderly subjects [78–80]. More recently, a meta-analysis of these and other studies in a total of 1972 hip fractured patients aged 78–84 demonstrated that vitamin D deficiency was associated with poorer quality of life and functional ability after fracture, while it did not influence the walking ability and length of stay in the hospital [74].

Retrospective and prospective studies reported inconsistent results as far as the possible role of pre-surgical vitamin D deficiency on mortality, as evaluated from 30 days to 3 years after hip fracture surgery [81]. Llombart et al. performed a meta-analysis of 9 cohort studies in 4409 patients with hip fracture aged 74–87 and concluded that reduced 25(OH)D levels were not associated with overall mortality when adjusting for confounders, such as age or comorbidities. Nevertheless, vitamin D insufficiency [defined as serum 25(OH)D between 20–29.9 ng/mL] and severe vitamin D deficiency [serum 25(OH)D < 10 ng/mL] significantly increased mortality after a year and 60 days from hip fracture surgery, respectively (Table 1) [75]. Wang et al. recently demonstrated significant associations between vitamin D deficiency and mortality but not with functional recovery in hip fractured subjects in a meta-analysis of 18 prospective and retrospective cohort studies (Table 1) [76].

Two recent studies focused on the possible role of vitamin D metabolites and vitamin D binding protein (DBP) on hip fracture risk and post-surgical outcomes. In a post hoc analysis of the Multi-Ethnic Study of Atherosclerosis, Hsu et al. used a composite of first hip and vertebral fracture as the primary outcome and found that total 25(OH)D, PTH, FGF23, and 24,25-dihydroxyvitamin D_3 to 25-hydroxyvitamin D_3 ratio were not independently associated with fracture risk in community-dwelling adults [82]. Authors did not report separate analysis for hip fracture in the study and acknowledged that the study was conducted in a population with low fracture risk and high percentage of vitamin D-replete individuals [82]. The post hoc analysis of the FOCUS trial reported significant associations between higher pre-surgical serum DBP levels and better 30- and 60-day mobility and lower 60-day mortality in a total of 260 hip fractured patients with a mean age of 81 [83].

In view of data gleaned from observational studies, several intervention studies including RCTs and testing the effect of vitamin D supplementation on hip fracture risk were conducted in different age groups and clinical settings. Owing to these discrepancies, as well as the heterogeneity in vitamin D dose regimens, and, most importantly, in the inclusion of a combination of calcium and vitamin D in the intervention group, results are not consistent. In this context, even results from meta-analyses of these studies failed to find agreement. Table 2 summarizes the results of the meta-analyses published in the last ten years (2014–2023) and including RCTs and non-RCTs studies that evaluated the effect of calcium combined with vitamin D on the risk for hip fractures [73,84–92].

Table 2. Summary of the meta-analyses of randomized controlled trials and clinical studies assessing the effect of the combination of calcium (Ca) and vitamin D (VitD) on hip fracture risk published from 2014 to 2023.

Author and Year of Publication	Design of Included Studies	Population	Setting	Intervention and Comparison	Results
Avenell et al., 2014 [84]	RCTs or quasi-randomized trials	post-menopausal women or men > 65	community or institutional	VitD or VitD-related compounds alone or in combination with Ca vs. placebo, no intervention, or Ca	- Analysis vs. placebo or no treatment (9 trials, 49,853 participants): significant reduction of hip fracture in the Ca/VitD group (RR 0.84, 95% CI 0.74–0.96) - Analysis vs. Ca (7 trials, 7411 participants): no significant reduction of hip fracture in the Ca/VitD group (RR 0.84, 95% CI 0.63–1.13)
Weaver et al., 2016 [85]	RCTs	generally healthy adults	community or institutional	Ca/VitD vs. placebo	- Analysis vs. placebo (6 trials): significant reduction of hip fracture in the Ca/VitD group (SRRE 0.7, 95% CI 0.56–0.87)
Zhao et al., 2017 [86]	RCTs	adults > 50	community	Ca, VitD and Ca/VitD vs. placebo or no treatment	- Analysis vs. placebo or no treatment (7 trials, 8657 participants): no significant reduction of hip fracture in the Ca/VitD group (RR 1.09, 95% CI 0.85–1.39)
Hu et al., 2019 [87]	RCTs	adults > 50	community	Ca, VitD and Ca/VitD vs. placebo	- Analysis vs. placebo: no significant reduction of hip fracture in the Ca/VitD group
Barrionuevo et al., 2019 [88]	RCTs	post-menopausal women with primary osteoporosis or osteopenia at risk for fragility fracture	not specified	various therapies (including Ca/VitD) vs. placebo; head-to-head comparisons	- Analysis vs. placebo: significant reduction of hip fracture in the Ca/VitD group (RR 0.81, 95% CI 0.71–0.93)
Yao et al., 2019 [73]	RCTs	adults	community or institutional	VitD and Ca/VitD vs. placebo or no treatment	- Analysis vs. placebo or no treatment (6 trials, 49,282 participants): significant reduction of hip fracture in the Ca/VitD group (RR 0.61, 95% CI 0.4–0.92)
Eleni et al., 2020 [89]	randomized trials	adults ≥ 50	not specified	Ca/VitD vs. placebo or no treatment	- Analysis vs. placebo or no treatment (8 trials, 68,957 participants): significant reduction of hip fracture in the Ca/VitD group (RR 0.84, 95% CI 0.72–0.97)

Table 2. *Cont.*

Author and Year of Publication	Design of Included Studies	Population	Setting	Intervention and Comparison	Results
Liu et al., 2020 [90]	RCTs	post-menopausal women	not specified	Ca/VitD vs. placebo or no treatment	- Analysis vs. placebo or no treatment (8 trials): significant reduction of hip fracture in the Ca/VitD group (RR 0.86, 95% CI 0.76–0.98)
Manoj et al., 2020 [91]	RCTs	women and men > 65	community or institutional	Ca/VitD vs. placebo or no treatment	- Analysis vs. placebo or no treatment (7 trials, 12,620 participants): significant reduction of hip fracture in the Ca/VitD group (OR 0.75, 95% CI 0.64–0.87)
Khatri et al., 2023 [92]	RCTs	adults ≥ 50 with fracture history	not specified	Ca/VitD vs. placebo or no treatment	- Analysis vs. placebo or no treatment (6 trials, 17,538 participants): no significant reduction of hip fracture in the Ca/VitD group (OR 1.1, 95% CI 0.86–1.4)

RCTs, randomized controlled trials; RR, risk ratio; SRRE, summary relative risk estimate; OR, odds ratio.

The meta-analysis by Yao et al. included 11 RCTs in which supplementation with various dosing regimens of vitamin D (ergo- and cholecalciferol) alone was employed to tests the effects on fracture risk and compared to placebo or no treatment [73]. Authors included RCTs performed in community dwelling as well as in institutionalized subjects [73]. In a total of 34,243 participants aged 66–85 and 740 hip fracture events, no association was detected between vitamin D supplementation and reduction in hip fracture [73]. The same authors performed a meta-analysis of 6 RCTs including 49,282 subjects with a mean age of 66 (730 hip fracture events) treated with daily 400–800 IU of vitamin D and 800–1200 mg of calcium for 6 years on average and demonstrated opposite results [73]. The intervention was indeed associated with a 16% reduction in hip fracture risk compared to placebo or no intervention [73]. Importantly, authors found consistent results between reduction in fracture risk reported in association with serum 25(OH)D levels by observational studies and those observed in the RCTs employing calcium and vitamin D [73]. Manoj et al. specifically meta-analyzed data from seven RCTs performed in subjects older than 65 and supplemented with daily 700–800 IU of cholecalciferol and 500–1200 mg of various calcium compounds [91]. In a total of 12,620 community dwelling and nursing home resident participants, authors reported roughly 25% reduction in hip fracture risk in the calcium and vitamin D combination group compared to placebo or no supplementation [91]. Risk reduction for hip fracture was as high as 30% in association with calcium and vitamin D supplementation when five RCTs performed only in older women were analyzed [91]. The analysis of three studies assessing femoral neck BMD in 483 subjects demonstrated no significant effect of calcium and vitamin D on this specific endpoint [91].

On the other side, the ancillary study of the Vitamin D and Omega-3 Trial (VITAL) failed to find significant effect of vitamin D on fracture risk in midlife and older individuals [93]. The study was conducted in 25,871 healthy adults aged 50 and older of whom only 12.9% had baseline serum 25(OH)D levels < 20 ng/mL and who were not selected for having osteoporosis [93]. Participants were followed up for a median of 5 years, and the daily dose of 2000 IU of cholecalciferol was employed. Compared to placebo, supplementation with cholecalciferol was not associated with any reduction in the incidence of first hip fracture occurring in 56 and 57 subjects in the placebo and vitamin D group, respectively [93]. Similarly to the VITAL, the DO-HEALTH trial reported that the primary prevention through vitamin D supplementation in 1900 healthy, community dwelling subjects older than 70 is not an effective strategy to be pursued to significantly reduce risk for nonvertebral fracture [94,95]. The recent meta-analysis of 18 RCTs including 39,759 subjects older than 50 and with history of fracture by Khatri et al. showed similar results [92]. Besides the observation of no significant effect of vitamin D supplementation alone on fracture risk, the authors reported also no association between calcium and vitamin D intake and reduction of hip fracture (Table 2) [92]. The results were confirmed in the subgroup analysis performed by calcium and vitamin D doses, gender, and baseline serum 25(OH)D levels [92].

In the attempt of reducing confusion generated by inconsistent reports published in the last decade on this sensitive topic, an umbrella review of systematic reviews and meta-analyses of RCTs was performed by Chakhtoura et al. [96]. The authors included 12 systematic reviews and meta-analyses assessing the effect of combined calcium and vitamin D on hip fracture risk and 15 on the effect of vitamin D alone. Though the studies by Leboff et al. were not included in the analysis, authors raised important points that collectively confirmed the absence of efficacy of vitamin D supplementation alone on fracture risk; combination of calcium and vitamin D was instead found to be effective for hip fracture risk reduction [96]. Eight systematic reviews and meta-analyses including community dwelling and institutionalized subjects treated with different doses of calcium and vitamin D showed efficacy of the combination on reduction of hip fracture [96]. Two systematic reviews and meta-analyses conducted in community dwelling subjects reported no effect of calcium and vitamin D [96]. The subgroup analysis demonstrated that factors potentially predicting a better response could be age, institutionalization, and baseline serum 25(OH)D < 20 ng/mL, but definitely need further confirmation by specifically designed studies [96].

5. Calcium Intake and Hip Fractures

Recovering from a hip fracture requires not just surgical intervention and physical therapy, but also nutritional support to aid the healing process. Calcium plays a crucial role in this recovery phase due to its importance in bone health.

After a hip fracture, the body's demand for calcium increases to support the repair of bone tissue. Calcium is a key building block for bones; it helps to restore bone density and strength, which is essential for the healing process. Adequate calcium intake can accelerate recovery, reduce the risk of a subsequent fracture, and support overall skeletal health.

The recommended daily calcium intake for older adults is about 1200 mg per day, according to the International Osteoporosis Foundation (IOF), Institute of Medicine and other referral institutions [97,98]. This can be achieved through diet, supplements, or a combination of both. Foods rich in calcium classically include dairy products (such as milk, cheese, and yogurt), green leafy vegetables (like kale and broccoli), fish with edible bones (such as sardines and canned salmon), and fortified foods (like certain cereals and plant-based milk). Chronic administration of drugs such as glucocorticoids or age-related malabsorption decreases intestinal calcium absorption, therefore increasing daily requirement of calcium achievable through diet intervention or supplements.

Although some metanalyses including 33 RCTs involving 51,145 participants, have shown that the increased intake of calcium and/or vitamin D per se is not associated with a lower risk of fragility fractures and hip fractures in particular [86,99], other recent longitudinal studies have shown that consumption of dairy foods, providing proteins and also minerals other than calcium, is protective against fractures. In two large U.S. cohorts, the Nurses' Health Study (NHS) of women and the Health Professionals Follow-up Study (HPFS) of men, one additional daily serving of total dairy, of which milk contributed about half, was found to be associated with a statistically significant 6% decreased risk of hip fracture among postmenopausal females and men [100]. A more recent analysis of the NHS cohort has further demonstrated that consumption of more than 2 daily servings of total dairy was associated with a decreased fracture risk, including hip fracture, as compared to 1 daily serving (hazard ratio [HR]: 0.74; 95% confidence interval [CI]: 0.61, 0.89) [101].

Older individuals are definitely at major risk to have low calcium and protein intake. One recent study examined the impact of dietary sources of calcium and protein on hip fractures and falls in older adults within residential care settings. This cluster RCT focused on increasing dairy food provision as a strategy to enhance calcium and protein intake aiming to reduce fracture and falls [44]. Intervention facilities were assisted by dieticians to increase dairy foods in meals and snacks, with methods such as fortifying milk used in recipes and offering dairy-based desserts. In this study 30 residential care facilities were randomized to provide each individual additional milk, yogurt and cheese to reach a total of 1142 mg calcium/day plus 69 f/day protein calcium, with results on the rate of falls and fractures as compared to care facilities maintaining their usual menu (700 mg/day calcium and 58 g/day protein) [44]. The intervention rapidly reduced the rate of total fractures by 33% and 46% of hip fracture in particular, along with 11% reduction of falls just in the first 5–3 months, demonstrating the effectiveness, promptness and easiness of increasing calcium and protein through dairy foods in preventing falls and fractures in older care residents [44]. An additional paper demonstrated cost-effectiveness of such intervention estimating costs related to fracture management (including ambulance, hospital, rehabilitation and residential care costs) [102].

Total calcium intake of people with fragility fractures is often low, especially in elderly people experiencing a hip fracture, and requires a prompt correction before the commencement of antifracture therapy. Moreover, defective intestinal absorption further decreases the already low calcium intake in the elderly.

Interventions to improve calcium intake and overall nutrition after hip fractures have been explored in various studies, highlighting the potential benefits for recovery and long-term bone health. A systematic review of RCTs evaluated post-surgery interventions for patients with hip fractures, including nutritional supplementation, among various other strategies such as rehabilitation, osteoporosis management, and fall prevention. This review highlighted the diversity of interventions tested and underscored the need for well-rounded postoperative care to improve outcomes in fragile hip fracture patients [103]. The included RCTs considered just calcium and vitamin D supplementation, while interventions to increase nutritional calcium intake by means of dairy products or other calcium sources were not mentioned [103]. Calcium supplements (i.e., calcium carbonate or citrate) along with vitamin D compounds are often easily prescribed, but they are often not tolerated and not taken in the long term.

Few experiences have been published about interventions to raise calcium intake, taking advantage of established FLSs. In a study from Taiwan, calcium intake (by supplements) in patients enrolled in an FLS, including a medication management service to improve medication adherence empowered by follow-up telephone interviews, was significantly improved, as well as vitamin D and protein intake, in the medium–long term, i.e., 1 year after the hip fracture event [53].

Tele-rehabilitation could support multicomponent interventions, such as nutritional intervention and physical exercise, after a hip fracture recovery. In this respect, ongoing RCTs, such as the ActiveFLS RCT, might give us evidence in the future that this approach could improve long-term nutritional status and physical performance, possibly decreasing falls and further fractures after a hip fracture.

Although it is impossible to dissect and isolate the effect of the improvement of calcium intake on the risk of further fractures, disability, and mortality, the optimization of calcium and vitamin D status is considered mandatory for each antifracture treatment that has to follow the fracture event.

6. Conclusions

The patient with a hip fracture has to be cared for with a multicomponent approach. The importance of nutrition, vitamin D status, and calcium intake is pivotal and should be regarded as a primary resource to be focused on. The target serum 25(OH)D of 30 ng/mL has to be reached and maintained in these patients as recommended by most guidelines [97,104], avoiding high doses, which are likely related to increased falls [105]. Vitamin D pro-hormones, such as cholecalciferol (vitamin D_3), ergocalciferol (vitamin D_2), or calcifediol (25(OH)D), can be effectively used to correct vitamin D status before and during antifracture therapy. Besides vitamin D status, which is usually optimized by supplements, calcium intake and overall nutrition should be evaluated and implemented in seniors after a hip fracture, also taking into account their generally suboptimal intestinal absorption. Too few short-term effective interventions have been carried out overall. The medium- and long-term outcomes of persistent changes in nutrition have yet to be assessed.

FLSs, where applied, do not often tackle these issues systematically. Besides nurses and case managers in the FLS, who can be employed to test nutritional status, specific dietary suggestions have to be dispensed by medical personnel or nutrition specialists. By addressing impaired nutrition in elderly individuals, particularly those with hip fractures, healthcare providers can enhance recovery outcomes, reduce complications, and improve the overall quality of life for this vulnerable population. In parallel, the effect of pharmacological intervention to prevent further fractures will be enhanced (Figure 1).

The inclusion of calcium and vitamin D supplementation is considered a potential strategy for improving compromised fracture healing in osteoporotic patients. Clinical evidence suggests that addressing post-traumatic bone loss through such supplementation could be beneficial, especially in the context of the broader management of osteoporosis and fracture healing.

These studies collectively suggest that targeted nutritional interventions, including calcium and protein intake through diet or supplements, along with comprehensive rehabilitation programs, can significantly contribute to improved recovery and mobility after hip fractures. Such strategies, along with targeted physical exercise, are critical for addressing the multifaceted needs of patients recovering from hip fractures, aiming not only at immediate recovery but also at preventing future fractures and enhancing overall bone and skeletal muscle health.

Author Contributions: Conceptualization: L.C., G.B., C.C. (Carla Caffarelli), G.M., N.N., S.M., C.R. and C.C. (Cristiana Cipriani); methodology: L.C., G.B., C.R. and C.C. (Cristiana Cipriani); writing—original draft preparation: L.C., G.B. and C.C. (Cristiana Cipriani); writing—review and editing: L.C., G.B., C.C. (Carla Caffarelli), G.M., N.N., S.M., C.R. and C.C. (Cristiana Cipriani); supervision, L.C., C.R. and C.C. (Cristiana Cipriani); funding acquisition: L.C. All authors have read and agreed to the published version of the manuscript.

Funding: This paper was supported by PRIN 2020 B53C21000260001 by the Italian Ministry of Research to L.C.

Conflicts of Interest: C.L. received consulting fees from UCB Pharma and Savio Pharma, and speaking fees from Abiogen Pharma and Bruno Farmaceutici. N.N. received consulting fees from UCB Pharma, Theramex, Abiogen Pharma and Bruno Farmaceutici. The remaining authors declare that the research was conducted in the absence of any commercial or financial relationships that could be construed as a potential conflict of interest.

References

1. Collaborators, G.B.D.F. Global, regional, and national burden of bone fractures in 204 countries and territories, 1990–2019: A systematic analysis from the Global Burden of Disease Study 2019. *Lancet Healthy Longev.* **2021**, *2*, e580–e592. [CrossRef] [PubMed]
2. Sing, C.W.; Lin, T.C.; Bartholomew, S.; Bell, J.S.; Bennett, C.; Beyene, K.; Bosco-Levy, P.; Bradbury, B.D.; Chan, A.H.Y.; Chandran, M.; et al. Global Epidemiology of Hip Fractures: Secular Trends in Incidence Rate, Post-Fracture Treatment, and All-Cause Mortality. *J. Bone Miner. Res.* **2023**, *38*, 1064–1075. [CrossRef] [PubMed]
3. Johnell, O.; Kanis, J.A. An estimate of the worldwide prevalence and disability associated with osteoporotic fractures. *Osteoporos Int.* **2006**, *17*, 1726–1733. [CrossRef] [PubMed]

Schemitsch, E.; Adachi, J.D.; Brown, J.P.; Tarride, J.E.; Burke, N.; Oliveira, T.; Slatkovska, L. Hip fracture predicts subsequent hip fracture: A retrospective observational study to support a call to early hip fracture prevention efforts in post-fracture patients. *Osteoporos. Int.* **2022**, *33*, 113–122. [CrossRef] [PubMed]

Legrand, M.A.; Chapurlat, R. Imminent fracture risk. *Jt. Bone Spine* **2021**, *88*, 105105. [CrossRef] [PubMed]

Compston, J. Reducing the treatment gap in osteoporosis. *Lancet Diabetes Endocrinol.* **2020**, *8*, 7–9. [CrossRef] [PubMed]

Patel, S.; Daniels, N.F.; Lim, J.A.; Zhou, A.K.; Thahir, A.; Krkovic, M. The Importance of Fracture Liaison Services to the Healthcare System: A Review. *Curr. Rheumatol. Rev.* **2023**, *19*, 36–41. [CrossRef] [PubMed]

Ni Lochlainn, M.; Cox, N.J.; Wilson, T.; Hayhoe, R.P.G.; Ramsay, S.E.; Granic, A.; Isanejad, M.; Roberts, H.C.; Wilson, D.; Welch, C.; et al. Nutrition and Frailty: Opportunities for Prevention and Treatment. *Nutrients* **2021**, *13*, 2349. [CrossRef] [PubMed]

Rizzoli, R.; Biver, E.; Brennan-Speranza, T.C. Nutritional intake and bone health. *Lancet Diabetes Endocrinol.* **2021**, *9*, 606–621. [CrossRef]

0. Kostecka, M.; Bojanowska, M. An evaluation of the nutritional status of elderly with the use of the MNA questionnaire and determination of factors contributing to malnutrition. A pilot study. *Rocz. Panstw. Zakl. Hig.* **2021**, *72*, 175–183. [CrossRef]

1. Sura, L.; Madhavan, A.; Carnaby, G.; Crary, M.A. Dysphagia in the elderly: Management and nutritional considerations. *Clin. Interv. Aging* **2012**, *7*, 287–298. [CrossRef] [PubMed]

2. Krishnamoorthy, Y.; Vijayageetha, M.; Kumar, S.G.; Rajaa, S.; Rehman, T. Prevalence of malnutrition and its associated factors among elderly population in rural Puducherry using mini-nutritional assessment questionnaire. *J. Family Med. Prim. Care* **2018**, *7*, 1429–1433. [CrossRef] [PubMed]

3. Bonjour, J.P. Dietary protein: An essential nutrient for bone health. *J. Am. Coll. Nutr.* **2005**, *24*, 526S–536S. [CrossRef] [PubMed]

4. Deutz, N.E.; Bauer, J.M.; Barazzoni, R.; Biolo, G.; Boirie, Y.; Bosy-Westphal, A.; Cederholm, T.; Cruz-Jentoft, A.; Krznaric, Z.; Nair, K.S.; et al. Protein intake and exercise for optimal muscle function with aging: Recommendations from the ESPEN Expert Group. *Clin. Nutr.* **2014**, *33*, 929–936. [CrossRef]

5. Wolfe, R.R. The role of dietary protein in optimizing muscle mass, function and health outcomes in older individuals. *Br. J. Nutr.* **2012**, *108* (Suppl. S2), S88–S93. [CrossRef] [PubMed]

6. Houston, D.K.; Nicklas, B.J.; Ding, J.; Harris, T.B.; Tylavsky, F.A.; Newman, A.B.; Lee, J.S.; Sahyoun, N.R.; Visser, M.; Kritchevsky, S.B.; et al. Dietary protein intake is associated with lean mass change in older, community-dwelling adults: The Health, Aging, and Body Composition (Health ABC) Study. *Am. J. Clin. Nutr.* **2008**, *87*, 150–155. [CrossRef] [PubMed]

7. Kerstetter, J.E.; O'Brien, K.O.; Caseria, D.M.; Wall, D.E.; Insogna, K.L. The impact of dietary protein on calcium absorption and kinetic measures of bone turnover in women. *J. Clin. Endocrinol. Metab.* **2005**, *90*, 26–31. [CrossRef] [PubMed]

8. Jehu, D.A.; Davis, J.C.; Falck, R.S.; Bennett, K.J.; Tai, D.; Souza, M.F.; Cavalcante, B.R.; Zhao, M.; Liu-Ambrose, T. Risk factors for recurrent falls in older adults: A systematic review with meta-analysis. *Maturitas* **2021**, *144*, 23–28. [CrossRef] [PubMed]

9. Verlaan, S.; Ligthart-Melis, G.C.; Wijers, S.L.J.; Cederholm, T.; Maier, A.B.; de van der Schueren, M.A.E. High Prevalence of Physical Frailty Among Community-Dwelling Malnourished Older Adults-A Systematic Review and Meta-Analysis. *J. Am. Med. Dir. Assoc.* **2017**, *18*, 374–382. [CrossRef]

0. Covinsky, K.E.; Martin, G.E.; Beyth, R.J.; Justice, A.C.; Sehgal, A.R.; Landefeld, C.S. The relationship between clinical assessments of nutritional status and adverse outcomes in older hospitalized medical patients. *J. Am. Geriatr. Soc.* **1999**, *47*, 532–538. [CrossRef]

1. Movassagh, E.Z.; Vatanparast, H. Current Evidence on the Association of Dietary Patterns and Bone Health: A Scoping Review. *Adv. Nutr.* **2017**, *8*, 1–16. [CrossRef] [PubMed]

2. Yaegashi, Y.; Onoda, T.; Tanno, K.; Kuribayashi, T.; Sakata, K.; Orimo, H. Association of hip fracture incidence and intake of calcium, magnesium, vitamin D, and vitamin K. *Eur. J. Epidemiol.* **2008**, *23*, 219–225. [CrossRef] [PubMed]

3. Fung, T.T.; Meyer, H.E.; Willett, W.C.; Feskanich, D. Association between Diet Quality Scores and Risk of Hip Fracture in Postmenopausal Women and Men Aged 50 Years and Older. *J. Acad. Nutr. Diet.* **2018**, *118*, 2269–2279.E4. [CrossRef] [PubMed]

4. Warensjo Lemming, E.; Byberg, L.; Hoijer, J.; Larsson, S.C.; Wolk, A.; Michaelsson, K. Combinations of dietary calcium intake and mediterranean-style diet on risk of hip fracture: A longitudinal cohort study of 82,000 women and men. *Clin. Nutr.* **2021**, *40*, 4161–4170. [CrossRef] [PubMed]

5. Zeng, F.F.; Xue, W.Q.; Cao, W.T.; Wu, B.H.; Xie, H.L.; Fan, F.; Zhu, H.L.; Chen, Y.M. Diet-quality scores and risk of hip fractures in elderly urban Chinese in Guangdong, China: A case-control study. *Osteoporos. Int.* **2014**, *25*, 2131–2141. [CrossRef] [PubMed]

6. Benetou, V.; Orfanos, P.; Feskanich, D.; Michaelsson, K.; Pettersson-Kymmer, U.; Byberg, L.; Eriksson, S.; Grodstein, F.; Wolk, A.; Jankovic, N.; et al. Mediterranean diet and hip fracture incidence among older adults: The CHANCES project. *Osteoporos. Int.* **2018**, *29*, 1591–1599. [CrossRef] [PubMed]

7. Langsetmo, L.; Hanley, D.A.; Prior, J.C.; Barr, S.I.; Anastassiades, T.; Towheed, T.; Goltzman, D.; Morin, S.; Poliquin, S.; Kreiger, N.; et al. Dietary patterns and incident low-trauma fractures in postmenopausal women and men aged >/= 50 y: A population-based cohort study. *Am. J. Clin. Nutr.* **2011**, *93*, 192–199. [CrossRef] [PubMed]

8. Shivappa, N.; Wirth, M.D.; Hurley, T.G.; Hebert, J.R. Association between the dietary inflammatory index (DII) and telomere length and C-reactive protein from the National Health and Nutrition Examination Survey-1999-2002. *Mol. Nutr. Food Res.* **2017**, *61*, 1600630. [CrossRef] [PubMed]

9. Zhang, Z.Q.; Cao, W.T.; Shivappa, N.; Hebert, J.R.; Li, B.L.; He, J.; Tang, X.Y.; Liang, Y.Y.; Chen, Y.M. Association Between Diet Inflammatory Index and Osteoporotic Hip Fracture in Elderly Chinese Population. *J. Am. Med. Dir. Assoc.* **2017**, *18*, 671–677. [CrossRef] [PubMed]

30. Zoltick, E.S.; Sahni, S.; McLean, R.R.; Quach, L.; Casey, V.A.; Hannan, M.T. Dietary protein intake and subsequent falls in old men and women: The Framingham Study. *J. Nutr. Health Aging* **2011**, *15*, 147–152. [CrossRef]
31. Misra, D.; Berry, S.D.; Broe, K.E.; McLean, R.R.; Cupples, L.A.; Tucker, K.L.; Kiel, D.P.; Hannan, M.T. Does dietary protein reduc hip fracture risk in elders? The Framingham Osteoporosis Study. *Osteoporos. Int.* **2011**, *22*, 345–349. [CrossRef] [PubMed]
32. Hayashi, A.P.; de Capitani, M.D.; Dias, S.F.; de Souza Goncalves, L.; Fernandes, A.L.; Jambassi-Filho, J.C.; de Santana, D.A Lixandrao, M.; Tavares Dos Santos Pereira, R.; Riani, L.; et al. Number of high-protein containing meals correlates with musc mass in pre-frail and frail elderly. *Eur. J. Clin. Nutr.* **2020**, *74*, 1047–1053. [CrossRef] [PubMed]
33. Webster, J.; Greenwood, D.C.; Cade, J.E. Foods, nutrients and hip fracture risk: A prospective study of middle-aged women. *Cli Nutr.* **2022**, *41*, 2825–2832. [CrossRef] [PubMed]
34. Webster, J.; Greenwood, D.C.; Cade, J.E. Risk of hip fracture in meat-eaters, pescatarians, and vegetarians: Results from the U Women's Cohort Study. *BMC Med.* **2022**, *20*, 275. [CrossRef] [PubMed]
35. Kondrup, J.; Allison, S.P.; Elia, M.; Vellas, B.; Plauth, M. ESPEN guidelines for nutrition screening 2002. *Clin. Nutr.* **2003**, *2* 415–421. [CrossRef]
36. Inciong, J.F.B.; Chaudhary, A.; Hsu, H.S.; Joshi, R.; Seo, J.M.; Trung, L.V.; Ungpinitpong, W.; Usman, N. Hospital malnutrition i northeast and southeast Asia: A systematic literature review. *Clin. Nutr. ESPEN* **2020**, *39*, 30–45. [CrossRef]
37. Han, T.S.; Yeong, K.; Lisk, R.; Fluck, D.; Fry, C.H. Prevalence and consequences of malnutrition and malnourishment in old individuals admitted to hospital with a hip fracture. *Eur. J. Clin. Nutr.* **2021**, *75*, 645–652. [CrossRef] [PubMed]
38. van Wissen, J.; van Stijn, M.F.; Doodeman, H.J.; Houdijk, A.P. Mini Nutritional Assessment and Mortality after Hip Fractu Surgery in the Elderly. *J. Nutr. Health Aging* **2016**, *20*, 964–968. [CrossRef] [PubMed]
39. Lumbers, M.; New, S.A.; Gibson, S.; Murphy, M.C. Nutritional status in elderly female hip fracture patients: Comparison with a age-matched home living group attending day centres. *Br. J. Nutr.* **2001**, *85*, 733–740. [CrossRef]
40. Kramer, I.F.; Blokhuis, T.J.; Verdijk, L.B.; van Loon, L.J.C.; Poeze, M. Perioperative nutritional supplementation and skelet muscle mass in older hip-fracture patients. *Nutr. Rev.* **2019**, *77*, 254–266. [CrossRef]
41. Shi, H.; Lu, J.H.; Wang, S.N.; Na, Q.; Xu, L.F.; Hong, J.A. Effect of early enteral nutrition in elderly patients with hip fractu during the perioperative period. *J. Back. Musculoskelet. Rehabil.* **2020**, *33*, 109–117. [CrossRef]
42. Sullivan, D.H.; Nelson, C.L.; Klimberg, V.S.; Bopp, M.M. Nightly enteral nutrition support of elderly hip fracture patients: A pil study. *J. Am. Coll. Nutr.* **2004**, *23*, 683–691. [CrossRef]
43. Wyers, C.E.; Reijven, P.L.M.; Breedveld-Peters, J.J.L.; Denissen, K.F.M.; Schotanus, M.G.M.; van Dongen, M.; Eussen, S.; Heyliger I.C.; van den Brandt, P.A.; Willems, P.C.; et al. Efficacy of Nutritional Intervention in Elderly After Hip Fracture: A Multicent Randomized Controlled Trial. *J. Gerontol. A Biol. Sci. Med. Sci.* **2018**, *73*, 1429–1437. [CrossRef]
44. Iuliano, S.; Poon, S.; Robbins, J.; Bui, M.; Wang, X.; De Groot, L.; Van Loan, M.; Zadeh, A.G.; Nguyen, T.; Seeman, E. Effect dietary sources of calcium and protein on hip fractures and falls in older adults in residential care: Cluster randomised controlle trial. *BMJ* **2021**, *375*, n2364. [CrossRef] [PubMed]
45. Inoue, T.; Maeda, K.; Nagano, A.; Shimizu, A.; Ueshima, J.; Murotani, K.; Sato, K.; Tsubaki, A. Undernutrition, Sarcopenia, an Frailty in Fragility Hip Fracture: Advanced Strategies for Improving Clinical Outcomes. *Nutrients* **2020**, *12*, 3743. [CrossRe [PubMed]
46. Van Stijn, M.F.; Bruins, A.A.; Vermeulen, M.A.; Witlox, J.; Teerlink, T.; Schoorl, M.G.; De Bandt, J.P.; Twisk, J.W.; Van Leeuwe P.A.; Houdijk, A.P. Effect of oral taurine on morbidity and mortality in elderly hip fracture patients: A randomized trial. *Int. Mol. Sci.* **2015**, *16*, 12288–12306. [CrossRef] [PubMed]
47. Bell, J.J.; Bauer, J.D.; Capra, S.; Pulle, R.C. Multidisciplinary, multi-modal nutritional care in acute hip fracture inpatients—Resul of a pragmatic intervention. *Clin. Nutr.* **2014**, *33*, 1101–1107. [CrossRef]
48. Duncan, D.G.; Beck, S.J.; Hood, K.; Johansen, A. Using dietetic assistants to improve the outcome of hip fracture: A randomise controlled trial of nutritional support in an acute trauma ward. *Age Ageing* **2006**, *35*, 148–153. [CrossRef]
49. Hoekstra, J.C.; Goosen, J.H.; de Wolf, G.S.; Verheyen, C.C. Effectiveness of multidisciplinary nutritional care on nutritional intak nutritional status and quality of life in patients with hip fractures: A controlled prospective cohort study. *Clin. Nutr.* **2011**, *3* 455–461. [CrossRef]
50. Invernizzi, M.; de Sire, A.; D'Andrea, F.; Carrera, D.; Reno, F.; Migliaccio, S.; Iolascon, G.; Cisari, C. Effects of essential amino aci supplementation and rehabilitation on functioning in hip fracture patients: A pilot randomized controlled trial. *Aging Clin. Ex Res.* **2019**, *31*, 1517–1524. [CrossRef]
51. Ooi, T.C.; Singh, D.K.A.; Shahar, S.; Rajab, N.F.; Vanoh, D.; Sharif, R.; Tan, M.P. Incidence and multidimensional predictors occasional and recurrent falls among Malaysian community-dwelling older persons. *BMC Geriatr.* **2021**, *21*, 154. [CrossRe [PubMed]
52. Valladales-Restrepo, L.F.; Castro-Osorio, E.E.; Ramirez-Osorio, J.; Echeverry-Martinez, L.F.; Sanchez-Rios, V.; Gaviria-Mendoz A.; Machado-Duque, M.E.; Machado-Alba, J.E. Characterization and effectiveness of a Fracture Liaison Services program i Colombia. *Arch. Osteoporos.* **2023**, *18*, 124. [CrossRef]
53. Chang, C.B.; Yang, R.S.; Chang, L.Y.; Peng, J.K.; Tsai, K.S.; Huang, W.J.; Yang, T.H.; Chan, D.C. One-year outcomes of a osteoporosis liaison services program initiated within a healthcare system. *Osteoporos. Int.* **2021**, *32*, 2163–2172. [CrossRe [PubMed]

54. Or, O.; Fisher Negev, T.; Hadad, V.; Shabtai, R.; Katzir, A.; Weil, Y.; Liebergall, M. Fracture Liaison Service for Hip Fractures: Is It A Game Changer? *Isr. Med. Assoc. J.* **2021**, *23*, 490–493.
55. Dempewolf, S.; Mouser, B.; Rupe, M.; Owen, E.C.; Reider, L.; Willey, M.C. What Are the Barriers to Incorporating Nutrition Interventions Into Care of Older Adults With Femoral Fragility Fractures? *Iowa Orthop. J.* **2023**, *43*, 172–182. [PubMed]
56. Awal, W.B.R.; Price, N.; Sadler, A.; Robinson, A.; Hymer, I.; Chen, J. Vitamin D deficiency in proximal femur fracture patients of South-East Queensland. *Australas. J. Aging* **2020**, *39*, e271–e277. [CrossRef] [PubMed]
57. Bakhtiyarova, S.; Lesnyak, O.; Kyznesova, N.; Blankenstein, M.A.; Lips, P. Vitamin D status among patients with hip fracture and elderly control subjects in Yekaterinburg, Russia. *Osteop. Int.* **2006**, *17*, 441–446. [CrossRef] [PubMed]
58. Bischoff-Ferrari, H.; Can, U.; Staehelin, H.; Platz, A.; Henschkowski, J.; Michel, B.; Dawson-Hughes, B.; Theiler, R. Severe vitamin D deficiency in Swiss hip fracture patients. *Bone* **2008**, *42*, 597–602. [CrossRef] [PubMed]
59. Moniz, C.; Dew, T.; Dixon, T. Prevalence of vitamin D inadequacy in osteoporotic hip fracture patients in London. *Curr. Med. Res. Opin.* **2005**, *21*, 1891–1894. [CrossRef]
60. Gallacher, S.J.; McQuillian, C.; Harkness, M.; Finlay, F.; Gallagher, A.P.; Dixon, T. Prevalence of vitamin D inadequacy in Scottish adults with non-vertebral fragility fractures. *Curr. Med. Res. Opin.* **2005**, *21*, 1355–1361. [CrossRef]
61. Bryson, D.J.; Neveda, J.; Ford, A.J.; Williams, S.C. The incidence of vitamin D deficiency amongst patients with a femoral neck fracture: Are current bone protection guidelines sufficient ? *Acta Orthop. Belg.* **2013**, *79*, 470–473. [PubMed]
62. Inderjeeth, C.A.; Barrett, T.; Al-Lahham, Y.; Mulford, J.; Nicklason, F.; Reberger, C. Seasonal variation, hip fracture and vitamin D levels in Southern Tasmania. *N. Z. Med. J.* **2002**, *115*, 183–185. [PubMed]
63. Dhanwal, D.K.; Samen, S.; Gautam, V.K.; Saha, R. Hip fracture patients in India have vitamin D deficiency and secondary hyperparathyroidism. *Osteop. Int.* **2013**, *24*, 553–557. [CrossRef] [PubMed]
64. Phusunti, S.; Suthutvoravut, W.; Unnanuntana, A.; Chotiyarnwong, P. The prevalence of hypovitaminosis D in patient with fragility hip fracture at a single institution in Thailand. *J. Med. Assoc. Thai.* **2016**, *99*, 1233–1238. [PubMed]
65. Farouk, O.; Mahran, D.G.; Said, H.G.; Alaa, M.M.; Eisa, A.A.; Said, G.Z.; Rashed, H.; Ez-Eldeen, A. Hypovitaminosis D among patients admitted with hip fracture to a level-1 Trauma Center in the sunny Upper Egypt: Prevalence and associated correlates. *Ger. Orthop. Surg. Rehab.* **2016**, *7*, 148–152. [CrossRef] [PubMed]
66. Lakkireddy, M.; Mudavath, S.V.; Karra, M.L.; Arora, A.J. Hypovitaminosis D in patients with osteoporotic hip fractures. *J. Clin. Orthop. Trauma.* **2019**, *10*, 768–773. [CrossRef] [PubMed]
67. Ramason, R.; Selvaganapathi, N.; Binte Ismail, N.H.; Wong, W.C.; Rajamoney, G.N.; Chong, M.S. Prevalence of vitamin D deficiency in patients with hip fracture seen in an Orthogeriatric Service in sunny Singapore. *Ger. Orthop. Surg. Rehab.* **2014**, *5*, 82–86. [CrossRef] [PubMed]
68. Lv, Q.B.; Gao, X.; Liu, X.; Shao, Z.X.; Xu, Q.H.; Tang, L.; Chi, Y.L.; Wu, A.M. The serum 25-hydroxyvitamin D levels and hip fracture risk: A meta-analysis of prospective cohort studies. *Oncotarget* **2017**, *8*, 39849–39858. [CrossRef] [PubMed]
69. Wang, N.; Chen, Y.; Ji, J.; Chang, J.; Yu, S.; Yu, B. The relationship between serum vitamin D and fracture risk in the elderly: A metaanalysis. *J. Orthop. Surg. Res.* **2020**, *15*, 1–10. [CrossRef]
70. Ghahfarrokhi, S.H.; Mohammadian-Hafshejani, A.; Sherwin, C.M.T.; Heidari-Soureshjani, S. Relationship between serum vitamin D and hip fracture in the elderly: A systematic review and meta-analysis. *J. Bone Miner. Metab.* **2022**, *40*, 541–553. [CrossRef]
71. Wanby, P.; Nobin, R.; Von, S.P.; Brudin, L.; Carlsson, M. Serum levels of the bone turnover markers dickkopf-1, sclerostin, osteoprotegerin, osteopontin, osteocalcin and 25-hydroxyvitamin D in Swedish geriatric patients aged 75 years or older with a fresh hip fracture and in healthy controls. *J. Endocrinol. Investig.* **2016**, *39*, 855–863. [CrossRef] [PubMed]
72. Feng, Y.; Cheng, G.; Wang, H.; Chen, B. The associations between serum 25-hydroxyvitamin D level and the risk of total fracture and hip fracture. *Osteoporos. Int.* **2017**, *28*, 1641–1652. [CrossRef] [PubMed]
73. Yao, P.; Bennett, D.; Mafham, M.; Lin, X.; Chen, Z.; Armitage, J.; Clarke, R. Vitamin D and calcium for the prevention of fracture. A systematic review and meta-analysis. *JAMA Network Open* **2019**, *2*, e1917789. [CrossRef]
74. Llombart, R.; Mariscal, G.; Barrios, C.; de la Rubia Ortí, J.E.; Llombart-Ais, R. Does vitamin D deficiency affect functional outcomes in hip fracture patients? A meta-analysis of cohort studies. *J. Endocrinol. Investig.* **2023**, 1–12. [CrossRef] [PubMed]
75. Llombart, R.; Mariscal, G.; Barrios, C.; de la Rubia Ortí, J.E.; Llombart-Ais, R. Impact of vitamin D deficiency on mortality in patients with hip fracture: A meta-analysis. *J. Am. Geriatr. Soc.* **2024**, *72*, 268–279. [CrossRef] [PubMed]
76. Wang, K.; Xia, C.; Zhou, L.; Zheng, Y.; Wang, X.; Cheng, L. The Association between Vitamin D Deficiency and the Risk of Mortality after Hip Fractures: A Systematic Review and Meta-Analysis. *J. Nutr. Sci. Vitaminol.* **2024**, *70*, 89–97. [CrossRef] [PubMed]
77. Hao, L.; Carson, J.L.; Schlussel, Y.; Noveck, H.; A Shapses, S. Vitamin D deficiency is associated with reduced mobility after hip fracture surgery: A prospective study. *Am. J. Clin. Nutr.* **2020**, *112*, 613–618. [CrossRef]
78. Sim, D.S.; Tay, K.; Howe, T.S.; Koh, S.B.J. Preoperative severe vitamin D deficiency is a significant independent risk factor for poorer functional outcome and quality of life 6 months after surgery for fragility hip fractures. *Osteop. Int.* **2021**, *32*, 2217–2224. [CrossRef] [PubMed]
79. Fu, G.; Wu, R.; Zhang, R.; Chen, D.; Li, H.; Zheng, Q.; Ma, Y. Preoperative vitamin D deficiency is associated with increased one-year mortality in Chinese geriatric hip fracture patients—A propensity score matching study. *Clin. Interv. Aging* **2023**, *18*, 263–272. [CrossRef]

80. Ingstad, F.; Solberg, L.B.; Nordsletten, L.; Thorsby, P.M.; Hestnes, I.; Frihagen, F. Vitamin D status and complications, readmissions and mortality after hip fracture. *Osteop. Int.* **2021**, *32*, 873–881. [CrossRef]
81. Nurmi-Lüthje, I.; Tiihonen, R.; Paattiniemi, E.-L.; Sarkkinen, H.; Naboulsi, H.; Pigg, S.; Kaukonen, J.-P.; Kataja, M.; Lüthje, P. Relatively low and moderate pre-fracture serum 25-hydroxyvitamin D levels associated with the highest survival in elderly hip fracture patients in Finland: A minimum 3-year follow-up. *Osteop. Int.* **2022**, *33*, 611–621. [CrossRef]
82. Hsu, S.; Criqui, M.H.; Ginsberg, C.; Hoofnagle, A.N.; Ix, J.H.; McClelland, R.L.; Michos, E.D.; Shea, S.J.; Siscovick, D.; Zelnick, L.R.; et al. Biomarkers of vitamin D metabolism and hip and vertebral fracture risk: The Multi-Ethnic Study of Atherosclerosis. *JBMR Plus* **2022**, *6*, e10697. [CrossRef] [PubMed]
83. Meng, L.; Wang, X.; Carson, J.L.; Schlussel, Y.; Shapses, S.A. Vitamin D binding protein and postsurgical outcomes and tissue injury markers after hip fracture: A prospective study. *J. Clin. Endocrinol. Metab.* **2024**, *109*, e18–e124. [CrossRef] [PubMed]
84. Avenell, A.; Mecheal, J.; O'Connell, D.L. Vitamin D and vitamin D analogues for preventing fractures in postmenopausal women and older men. *Cochrane Database Syst. Rev.* **2014**, *4*, CD000227. [CrossRef] [PubMed]
85. Weaver, C.M.; Anna, D.; Boushey, C.J.; Dawson-Hughes, B.; Lappe, J.M.; LeBoff, M.S.; Liu, S.; Looker, A.C.; Wallace, T.C.; Wang, D.D. Calcium plus vitamin D supplementation and risk of fractures: An updated meta-analysis from the National Osteoporosis Foundation. *Osteop. Int.* **2016**, *27*, 367–376. [CrossRef]
86. Zhao, J.G.; Zeng, X.T.; Wang, J.; Liu, L. Association Between Calcium or Vitamin D Supplementation and Fracture Incidence in Community-Dwelling Older Adults: A Systematic Review and Meta-analysis. *JAMA* **2017**, *318*, 2466–2482. [CrossRef] [PubMed]
87. Hu, Z.C.; Tang, Q.; Sang, C.M.; Tang, L.; Li, X.; Zheng, G.; Feng, Z.H.; Xuan, J.W.; Shen, Z.H.; Shen, L.-Y.; et al. Comparison of fracture risk using different supplemental doses of vitamin D, calcium or their combination: A network meta-analysis of randomised controlled trials. *BMJ Open* **2019**, *9*, e024595. [CrossRef] [PubMed]
88. Barrionuevo, P.; Kapoor, E.; Asi, N.; Alahdab, F.; Mohammed, K.; Benkhadra, K.; Almasri, J.; Farah, W.; Sarigianni, M.; Muthusamy, K.; et al. Efficacy of pharmacological therapies for the prevention of fractures in postmenopausal women: A network meta-Analysis. *J. Clin. Endocrinol. Metab.* **2019**, *104*, 1623–1630. [CrossRef] [PubMed]
89. Elen, A.; Panagiotis, P. A systematic review and meta-analysis of vitamin D and calcium in preventing osteoporotic fractures. *Clin. Rheumatol.* **2020**, *39*, 3571–3579. [CrossRef]
90. Liu, C.; Kuang, X.; Li, K.; Guo, X.; Deng, Q.; Li, D. Effects of combined calcium and vitamin D supplementation on osteoporosis in postmenopausal women: A systematic review and meta-analysis of randomized controlled trials. *Food Funct.* **2020**, *11*, 10817. [CrossRef]
91. Manoj, P.; Derwin, R.; George, S. What is the impact of daily oral supplementation of vitamin D3 (cholecalciferol) plus calcium on the incidence of hip fracture in older people? A systematic review and meta-analysis. *Int. J. Older People Nurs.* **2023**, *18*, e12492. [CrossRef] [PubMed]
92. Khatri, K.; Kaur, M.; Dhir, T.; Kankaria, A.; Arora, H. Role of calcium &/or vitamin D supplementation in preventing osteoporotic fracture in the elderly: A systematic review & meta-analysis. *Indian. J. Med. Res.* **2023**, *158*, 5–16. [CrossRef] [PubMed]
93. LeBoff, M.S.; Chou, S.H.; Ratliff, K.A.; Cook, N.R.; Khurana, B.; Kim, E.; Cawthon, P.M.; Bauer, D.C.; Black, D.; Gallagher, J.C.; et al. Supplemental vitamin D and incident fractures in midlife and older adults. *N. Engl. J. Med.* **2022**, *387*, 299–309. [CrossRef]
94. Bischoff-Ferrari, H.A.; Vellas, B.; Rizzoli, R.; Kressig, R.W.; da Silva, J.A.P.; Blauth, M.; Felson, D.T.; McCloskey, E.V.; Watzl, B.; Hofbauer, L.C.; et al. Effect of vitamin D supplementation, omega-3 fatty acid supplementation, or a strength-training exercise program on clinical outcomes in older adults: The DO-HEALTH randomized clinical trial. *JAMA* **2020**, *324*, 1855–1868. [CrossRef] [PubMed]
95. LeBoff, M.S.; Bischoff-Ferrari, H.A. The effects of vitamin D supplementation on musculoskeletal health: The VITAL and DO-Health Trials. *J. Gerontol. A Biol. Sci. Med. Sci.* **2023**, *78*, S73–S78. [CrossRef] [PubMed]
96. Chakhtoura, M.; Bacha, D.S.; Gharios, C.; Ajjour, S.; Assaad, M.; Jabbour, Y.; Kahale, F.; Bassatne, A.; Antoun, S.; Akl, E.A.; et al. Vitamin D supplementation and fractures in adults: A systematic umbrella review of meta-analyses of controlled trials. *J. Clin. Endocrinol. Metab.* **2022**, *107*, 882–898. [CrossRef] [PubMed]
97. Kanis, J.A.; Cooper, C.; Rizzoli, R.; Reginster, J.Y. Scientific Advisory Board of the European Society for Clinical and Economic Aspects of Osteoporosis and Osteoarthritis (ESCEO) and the Committees of Scientific Advisors and National Societies of the International Osteoporosis Foundation (IOF). Executive summary of European guidance for the diagnosis and management of osteoporosis in postmenopausal women. *Aging Clin. Exp. Res.* **2019**, *31*, 15–17. [CrossRef] [PubMed]
98. Optimal calcium intake. *NIH Consens Statement* **1994**, *12*, 1–31.
99. Bolland, M.J.; Leung, W.; Tai, V.; Bastin, S.; Gamble, G.D.; Grey, A.; Reid, I.R. Calcium intake and risk of fracture: Systematic review. *BMJ* **2015**, *351*, h4580. [CrossRef]
100. Feskanich, D.; Meyer, H.E.; Fung, T.T.; Bischoff-Ferrari, H.A.; Willett, W.C. Milk and other dairy foods and risk of hip fracture in men and women. *Osteoporos. Int.* **2018**, *29*, 385–396. [CrossRef]
101. Yuan, M.; Hu, F.B.; Li, Y.; Cabral, H.J.; Das, S.K.; Deeney, J.T.; Zhou, X.; Paik, J.M.; Moore, L.L. Types of dairy foods and risk of fragility fracture in the prospective Nurses' Health Study cohort. *Am. J. Clin. Nutr.* **2023**, *118*, 1172–1181. [CrossRef] [PubMed]
102. Baek, Y.; Iuliano, S.; Robbins, J.; Poon, S.; Seeman, E.; Ademi, Z. Reducing hip and non-vertebral fractures in institutionalised older adults by restoring inadequate intakes of protein and calcium is cost-saving. *Age Ageing* **2023**, *52*, afad114. [CrossRef] [PubMed]

103. Phang, J.K.; Lim, Z.Y.; Yee, W.Q.; Tan, C.Y.F.; Kwan, Y.H.; Low, L.L. Post-surgery interventions for hip fracture: A systematic review of randomized controlled trials. *BMC Musculoskelet. Disord.* **2023**, *24*, 417. [CrossRef] [PubMed]
104. Sanders, K.M.; Seibel, M.J. Therapy: New findings on vitamin D3 supplementation and falls—When more is perhaps not better. *Nat. Rev. Endocrinol.* **2016**, *12*, 190–191. [CrossRef]
105. Bischoff-Ferrari, H.A.; Dawson-Hughes, B.; Orav, E.J.; Staehelin, H.B.; Meyer, O.W.; Theiler, R.; Dick, W.; Willett, W.C.; Egli, A. Monthly High-Dose Vitamin D Treatment for the Prevention of Functional Decline: A Randomized Clinical Trial. *JAMA Intern. Med.* **2016**, *176*, 175–183. [CrossRef]

Disclaimer/Publisher's Note: The statements, opinions and data contained in all publications are solely those of the individual author(s) and contributor(s) and not of MDPI and/or the editor(s). MDPI and/or the editor(s) disclaim responsibility for any injury to people or property resulting from any ideas, methods, instructions or products referred to in the content.

MDPI AG
Grosspeteranlage 5
4052 Basel
Switzerland
Tel.: +41 61 683 77 34

Nutrients Editorial Office
E-mail: nutrients@mdpi.com
www.mdpi.com/journal/nutrients

Disclaimer/Publisher's Note: The title and front matter of this reprint are at the discretion of the Guest Editors. The publisher is not responsible for their content or any associated concerns. The statements, opinions and data contained in all individual articles are solely those of the individual Editors and contributors and not of MDPI. MDPI disclaims responsibility for any injury to people or property resulting from any ideas, methods, instructions or products referred to in the content.

www.ingramcontent.com/pod-product-compliance
Lightning Source LLC
Chambersburg PA
CBHW062057220926
51850CB00010B/2798
* 9 7 8 3 7 2 5 8 6 7 4 8 6 *